“Like almost all the church fathers, [illegible] rightly saw as the key to the Christian worldview. Dr. Ortlund takes us back to the man and his beliefs, at once so distant from and yet so near to our own concerns. Modern readers will be challenged by Augustine’s insights, and by entering into dialogue with him, they may find answers to the dilemmas they confront. An exciting book on a key topic for our times.”

Gerald Bray, research professor of divinity, Beeson Divinity School at Samford University

“We need pastors like Gavin Ortlund, and we need books written by pastors like Gavin Ortlund! His opening chapter on humility sets the stage for a book that is contextually responsible, academically sound, and pastorally motivated. I highly recommend this book as a rewarding and promising retrieval of Augustine’s doctrine of creation for the good of the church.”

Craig D. Allert, professor of religious studies, Trinity Western University

“As debates about creation, evolution, and the historical Adam come to a crucial new juncture among evangelicals today, I can hardly imagine a better discussion partner from the church’s tradition than Augustine, with his unwavering commitment to the truth of Scripture, his fearless willingness to pursue difficult questions, and his humble refusal to give rash and hasty answers. Gavin Ortlund gives us a well-rounded account of what Augustine’s exegesis of Genesis brings to the table.”

Phillip Cary, professor of philosophy at Eastern University

“This remarkable book offers a finely textured yet accessible interpretation of Augustine’s views on creation, at the same time relating his thought to contemporary issues in a way that is creative, responsible, and compelling. I commend this book with enthusiasm to any Christian in search of insight into debates about creation and science, to both scholars and students interested in Augustine’s thinking on creation, and to all those who seek a first-rate model of humble, rigorous, and faithful theological scholarship for the sake of the church.”

Han-luen Kantzer Komline, Western Theological Seminary

“What can the ancient bishop Augustine of Hippo contribute to contemporary debates regarding creation, the age of the earth, and evolution? A lot, as it turns out. Readers will find Gavin Ortlund’s masterful study of Augustine’s doctrine of creation to be a smart, humble, and immensely helpful exercise of theological reflection on a most vexing question.”

Scott Manetsch, editor of *The Reformation and the Irrepressible Word of God*, professor of church history at Trinity Evangelical Divinity School

"What do we who live in the post-industrial revolution twenty-first century have to learn about creation from a fifth-century North African bishop? As it turns out, quite a lot. First and foremost, Augustine helps us learn how to think, not only what to think. In *Retrieving Augustine's Doctrine of Creation*, Gavin Ortlund invites us into a conversation with one of the greatest minds of late antiquity to explore together the fundamental distinction between 'nature' and 'creation'; the former being the idolatrous attempt to perceive our reality as independent, while the latter restoring what Ortlund terms 'a holistic framework for how to live as God's creatures in God's world.' This is a book that needs to be read slowly, for neither the topic nor the transformative effect can be rushed. And there is no better interlocutor than Augustine to help us move from our autonomous, self-referential idolatry to the Creator of all, whose image we bear. Ortlund has done us a great favor. *Tolle, lege!*"

George Kalantzis, professor of theology and director of the Wheaton Center for Early Christian Studies

"People from all sides of the church's discussion on origins have cherry-picked quotations from Augustine to bolster their views, without digging in to his actual doctrine of creation. In so doing, we attempt to cast Augustine in our own image. Ortlund has done us all a service by presenting a much more comprehensive understanding of Augustine's thought on creation and retrieving his voice from across the centuries. I predict that Ortlund's treatment of Augustine will also be disappointing to different kinds of people—those looking to him merely to prop up their own theories. He gives us a more complex, sophisticated, and surprising Augustine—one that makes me want to read more Augustine . . . and more from Ortlund."

J. B. Stump, vice president of BioLogos

RETRIEVING AUGUSTINE'S DOCTRINE *of* CREATION

Ancient Wisdom for Current Controversy

Gavin Ortlund

An imprint of InterVarsity Press
Downers Grove, Illinois

InterVarsity Press
P.O. Box 1400, Downers Grove, IL 60515-1426
ivpress.com
email@ivpress.com

InterVarsity Press® is the book-publishing division of InterVarsity Christian Fellowship/USA®, a movement of students and faculty active on campus at hundreds of universities, colleges, and schools of nursing in the United States of America, and a member movement of the International Fellowship of Evangelical Students. For information about local and regional activities, visit intervarsity.org.

Cover design and image composite: Autumn Short
Interior design: Daniel van Loon
Images: beige screen pattern: © DavidMSchrader / iStock / Getty Images Plus
The Creation of the World: © The Creation of the World and the Expulsion from Paradise by Giovanni di Paolo di Grazia at the Metropolitan Museum of Art, New York, USA / Bridgeman Images
old paper: © pablohart / E+ / Getty Images
ornate vintage design: © Alex_Bond / iStock / Getty Images

ISBN 978-0-8308-5324-3 (print)
ISBN 978-0-8308-5325-0 (digital)

Printed in the United States of America ♾

InterVarsity Press is committed to ecological stewardship and to the conservation of natural resources in all our operations. This book was printed using sustainably sourced paper.

Library of Congress Cataloging-in-Publication Data
A catalog record for this book is available from the Library of Congress.

P 25 24 23 22 21 20 19 18 17 16 15 14 13 12 11 10 9 8 7 6 5 4 3
Y 41 40 39 38 37 36 35 34 33 32 31 30 29 28 27 26 25 24

FOR ERIC, KRISTA, AND DANE,

each of whom I admire

and whose friendship brightens

and enlarges life

CONTENTS

NOTE ON CITATIONS

Citations of Augustine's works are generally drawn from the critical edition of Augustine's Latin text in *Corpus scriptorum ecclesiasticorum Latinorum* (Vienna: Tempsk, 1894–1900) and are listed as follows: CSEL, volume:part, page number. When a particular work is not included in CSEL, I have occasionally referenced it in Jacques-Paul Migne's *Patrologia Latina* (Paris: 1844–1855) and listed as PL volume:column. Unless otherwise cited, for Augustine's commentaries on Genesis I have relied on the translations of Edmund Hill in the relevant volumes in The Works of Saint Augustine: A Translation for the 21st Century (Hyde Park, NY: New City, used with permission); for the *Confessiones*, I have used those of R. S. Pine-Coffin in Saint Augustine, *Confessions* (New York: Penguin, 1961); for *De civitate Dei*, I have used those of Marcus Dods in Saint Augustine, *The City of God* (Urbana, IL: Project Gutenberg, 2014, www.gutenberg.org/ebooks/45304.); and for *De libero arbitrio*, I have used those of Anna S. Benjamin and L. H. Hackstaff in *On Free Choice of the Will* (Upper Saddle River, NJ: Prentice Hall, 1964). Any exceptions to this practice are cited in the footnotes, as are references to Augustine's other works, letters, and sermons.

ACKNOWLEDGMENTS

THIS BOOK WAS COMPLETED during my resident fellowship during the 2017–2018 school year at Trinity Evangelical Divinity School, through a generous grant provided by the Templeton Religious Trust, as a part of *The Creation Project* overseen by the Carl F. H. Henry Center for Theological Understanding. I am very grateful to the Templeton Religious Trust for their generosity and vision, and to the Henry Center for their hospitality and support.

Our year in Chicago was a very fruitful season of learning, and over the course of the fellowship I benefited from interacting with a wide variety of people involved in the Creation Project. I enjoyed a weekly lunch discussion with the Henry Center staff and other fellows, including Tom McCall, Geoff Fulkerson, Joel Chopp, Dick Averbeck, Marc Cortez, Daniel Houck, Stephen Williams, and Nathan Chambers. In March 2018 we discussed an earlier draft of chapter 4, during which I received helpful feedback. At Dabar Conferences I especially benefited from interacting with William Lane Craig, J. Richard Middleton, Bill Kynes, A. J. Roberts, Fuz Rana, John Walton, John Hilbur, Jim Stump, and Paul Copan.

Starting during my years at Covenant Seminary and continuing through my time at the Creation Project, Jack Collins has contributed significantly to my understanding of the book of Genesis, and biblical hermeneutics more generally. I am grateful for his encouragement and all he has taught me. Matthew Levering was a great source of encouragement during this project, and went out of his way to give counsel and feedback about my work. During my time in Chicago I became friends with Todd Wilson, and my involvement with the St. Anselm Fellowship of the Center for Pastor Theologians has been an encouragement to me in my scholarship. Before and after our season in Chicago I spent time at Reasons to Believe as a visiting scholar, and I benefited much from my interactions there. I'm particularly grateful to Hugh Ross for his friendship and support.

An earlier version of chapter 4 was read as a part of the Colloquium on Creation and the Problem of Evil under the leadership of the Chicago Theological Institute, held at Wheaton College, March 23–24, 2018. Several paragraphs in the introduction and chapters one through three appeared earlier in a more popular-level form in the online periodical *Sapientia*.[1] The following friends were gracious enough to review and comment on an earlier draft of several chapters: Scott Manetsch, Daniel Houck, Austin Freeman, Dave Lauer, and Eric Ortlund. Scott Manetsch was a help and encouragement to me in my effort to publish the book. I enjoyed working with everyone at IVP, and in particular I am indebted to David McNutt for his openness to considering this project and his helpful collaboration along the way.

My wife Esther deserves thanks most of all—for being willing to take a break from pastoral ministry in Southern California to have an adventure in the Midwest, embrace a snowy Chicago winter, and parent several young children (one of whom arrived during the fellowship), and through it all never wavering in her friendship, support, and commitment. One thing is for sure: I am blessed beyond what I deserve.

[1]See Gavin Ortlund, "Can the Creation Debates Find Rest in Augustine?," *Sapientia*, August 14, 2017, https://henrycenter.tiu.edu/2017/08/can-the-creation-debates-find-rest-in-augustine; "What We Forget About Creation: How Augustine Expands Our Vision," *Sapientia*, August 21, 2017, https://henrycenter.tiu.edu/2017/08/what-we-forget-about-creation; "The Missing Virtue in the Creation Debates: Augustine on Why Humility Matters," *Sapientia*, August 28, 2017, https://henrycenter.tiu.edu/2017/08/the-missing-virtue-in-the-creation-debates; "Did Augustine Read Genesis 1 Literally?," *Sapientia*, September 4, 2017, https://henrycenter.tiu.edu/2017/09/did-augustine-read-genesis-1-literally.

INTRODUCTION

CAN CREATION DEBATES FIND THEIR REST IN AUGUSTINE?

Observe the beauty of the world, and praise the plan of the creator. Observe what he made, love the one who made it. Hold on to this maxim above all; love the one who made it, because he also made you, his lover, in his own image.

SERMON 68.5

IMAGINE A YOUNG MAN in his late teen years. He has recently moved to the city to go to school. In the course of his study, he becomes convinced that the Genesis creation account is inconsistent with the most sophisticated intellectual trends of the day. He rejects the Christian faith in which he was raised, giving his twenties to youthful sins and worldly ambition.

Eventually, he encounters Christians who hold to a different interpretation of the early chapters of Genesis, and his intellectual critique of Christianity is undermined. He enters into a time of indecision and deep angst. His mother continues to pray for him. Finally, after much personal struggle, he has a dramatic conversion experience.

This is the testimony of St. Augustine (354–430), who is arguably the greatest of the church fathers and the most influential Christian theologian in the history of the church.[1] But, in its broad outline, it is

[1]In his highly regarded history of Christian doctrine, Jaroslav Pelikan writes of Augustine that "there is probably no Christian theologian—Eastern or Western, ancient or medieval or modern, heretical or orthodox—whose historical influence can match his" (*The Christian Tradition: A*

a story that seems to replay itself again and again today. The details are different, of course. For instance, our threat today comes from naturalism, while Augustine's came from Manichaeism.[2] But the overall scenario is only too familiar to us—particularly because today, stories likes this often lack a happy ending.

CREATION WAS AT THE HEART OF AUGUSTINE'S THEOLOGY AND LIFE

Many people are unaware of the role that Genesis played in Augustine's conversion, and most would not identify the doctrine of creation as the high point in his theology. When we think of Augustine, we tend to think of his emphasis on divine grace, or his high doctrine of the church, or his penetrating insights into the Trinity.

But in many ways, the doctrine of creation is at the very heart of Augustine's Christian faith, his pastoral vocation, and the overall shape of his theology. Had Augustine not heard St. Ambrose preach allegorically on Genesis 1 in 384, he might never have come to fight Donatism or Pelagianism. More significantly, Augustine continued to wrestle with the doctrine of creation throughout his life, and it became an integral part of his theology. Not only did he write three distinct commentaries on Genesis, but the doctrine of creation comes up at pivotal moments throughout his other important works. For instance, the *Confessions* climax into an exegesis of Genesis 1, and *The City of God* grounds its vision of the two cities in a lengthy treatment of the doctrine of creation, particularly the Genesis account.[3]

History of the Development of Doctrine, vol. 1, *The Emergence of the Catholic Tradition 100–600* [Chicago: University of Chicago Press, 1971], 292).

[2]Manichaeism was a dualistic religion that combined Gnostic and pagan ideas with Christian ones. It was popular in fourth-century Carthage, where Augustine studied as a young man. Augustine became a Manichaean for almost ten years before converting back to Christianity. Much of his doctrine of creation is developed in response to alternative Manichaean theories. For a summary of Manichaean beliefs, see Gerald Bonner, *St. Augustine of Hippo: Life and Controversies*, 3rd ed. (Norwich: Canterbury, 2002), 157-92.

[3]The *City of God* was occasioned by the charge that the Christian faith was to blame for the sack of Rome. The first part of the book responds to this charge, while the second part extends this

In all of this, Augustine wrestled with the doctrine of creation at a profoundly existential level. At the risk of overstatement, I might suggest that creation was to Augustine what justification was to Luther, or divine transcendence was to Barth—the area of theology that, because of a theologian's own personal journey, comes to an especially vigorous expression and is visible in almost everything they wrote.

Because it was forged in the context of apologetic dispute, Augustine's doctrine of creation has a kind of philosophical flair, rather than being strictly exegetically or pastorally driven. Even his exegetical works devote considerable space and energy to digressions concerning the origin of the soul, or the nature of memory, or various Manichaean views. Thus, in engaging Augustine's doctrine of creation we gather a sense of how Christianity as a whole made sense to him—how it was better than its rival intellectual and religious systems at providing an explanation for the complexity of the world, the intensity of the contest between good and evil, and the strange longings of the human soul.

Augustine's doctrine of creation was recognized as authoritative in his own lifetime. When Jerome was asked a question about the origins of the soul, his response enumerated five possible answers—but he ultimately commended his reader to go to the "holy and learned man, bishop Augustine, who could teach you *viva voce*, as they say, and explain his own view and, in fact, our view by himself."[4] Jerome's own occasional differences with Augustine on this point make his deference to him in this context all the more revealing. In the modern era as well, interpreters of Augustine have detected the relevance of

---

discussion into a broader interpretation of history. This second part, initiated in book 11, begins with a treatment of creation. I will discuss the significance of creation in the *Confessions* in greater detail in chapter 1.

[4]Letter 165.1, in *Letters (Epistulae) 156–210*, ed. Boniface Ramsey, trans. Roland Teske, The Works of St. Augustine: A Translation for the 21st Century (Hyde Park, NY: New City Press, 2004), 74.

his work to science-faith issues. Galileo Galilei, for instance, quoted from Augustine's commentary on Genesis over ten times in defense of his theories.[5]

Nonetheless, for all its significance, the doctrine of creation is not one of the better-known aspects of Augustine's theology. As N. Joseph Torchia notes, "Despite the seeming inexhaustiveness of investigations into the life and work of St. Augustine of Hippo, his theology of creation remains a relatively neglected area of his thought."[6] Moreover, when Augustine's doctrine of creation is engaged, it is often pressed into the service of a particular contemporary concern or ideology.[7] Occasionally this leads to a certain distortion of Augustine's views, such that one can find his name simultaneously called down on opposite sides of various debates. Already in the 1920s Henry Woods could lament the association of Augustine's conception of *rationes seminales* with Darwin's theory of biological evolution,[8] and in recent years Alister McGrath has appropriated Augustine's thought to provide a theological framework for engaging modern science, especially evolutionary theory.[9] In the other direction, young-earth creationists balk at any association of Augustine with

---

[5]Note the discussion in Tarcisius van Bavel, "The Creator and the Integrity of Creation in the Fathers of the Church," *Augustinian Studies* 21 (1990): 1.

[6]N. Joseph Torchia, *Creatio Ex Nihilo and the Theology of Augustine: The Anti-Manichaean Polemic and Beyond*, American University Studies 205 (New York: Peter Lang, 1999), ix.

[7]Of course, there is nothing wrong with approaching Augustine in light of contemporary concerns and questions (such is the aim of this book). The danger is that in many such efforts, insufficient attention is given to Augustine's original context. For a more careful usage of Augustine in relation to a contemporary concern, see Scott A. Dunham, *The Trinity and Creation in Augustine: An Ecological Analysis* (Albany, NY: State University of New York Press, 2008).

[8]Henry Woods, *Augustine and Evolution: A Study in the Saint's* De Genesi ad litteram *and* De Trinitate (New York: The Universal Knowledge Foundation, 1924). Even earlier associations of *rationes seminales* with evolution are documented in the classic study by Eugène Portalié, *A Guide to the Thought of Saint Augustine*, trans. Ralph J. Bastian (1902; repr., Chicago: Henry Regnery, 1960), xxxi-xxxii, 139-41.

[9]Alister McGrath, *A Fine-Tuned Universe: The Quest for God in Science and Theology* (Louisville, KY: Westminster John Knox, 2009), 95-108; McGrath, *The Passionate Intellect: Christian Faith and the Discipleship of the Mind* (Downers Grove, IL: InterVarsity Press, 2010); McGrath, "Augustine's Origin of Species: How the Great Theologian Might Weigh in on the Darwin Debate," *Christianity Today* 53:5 (May 2009), www.christianitytoday.com/ct/2009/may/22.39.html.

the earth being millions of years old,[10] and are eager to claim his legacy as their own.[11]

In addition to these difficulties there stands the more general challenge of the persistence of certain unfavorable representations of Augustine's thought and personality in popular impression, and occasionally in academic portrait. Perhaps because his influence on Christianity and Western culture has been so far reaching, he is sometimes read through the lens of that (real or perceived) influence. His reputation has been accordingly controversial. Rowan Williams, for instance, references the wide array of "clichés about Augustine's alleged responsibility for Western Christianity's supposed obsession with the evils of bodily existence or sexuality, or its detachment from the world of public ethics, or its excessively philosophical understanding of God's unity, or whatever else is seen as the root of all theological evils."[12] Some of these views will be addressed throughout this book as they impinge materially on the topics covered, especially those that tend to border on caricature—for instance, Colin Gunton's critique of Augustine as the progenitor of all the dreadful monisms that have deteriorated modern Western civilization. For the greater part, I hope that simply making fresh contact with Augustine will weaken or overturn many of the various unhappy "Augustinianisms" that continuously crop up in historical texts. For those who have only encountered Augustine indirectly, the mere experience of a careful reading of his actual writings is often enough to temper the popular

---

[10]For instance, Louis Lavallee, "Augustine on the Creation Days," *Journal Of the Evangelical Theological Society* 32:4 (1989): 464, finds fault with "using Augustine's illustrious name to support harmonizing Genesis 1 and the idea of an ancient earth and/or evolutionary development." Lavallee cites this usage of Augustine in the writings of Charles Hodge, William Shedd, and James Orr. Lavellee's own engagement with Augustine consists primarily of block quotes with relatively little commentary and evaluation.

[11]E.g., James R. Mook, "The Church Fathers on Genesis, the Flood, and the Age of the Earth," in *Coming to Grips with Genesis: Biblical Authority and the Age of the Earth*, ed. Terry Mortenson and Thane H. Ury (Green Forest, AZ: Master Books, 2008), 35-38, 48; Terry Mortenson and A. Peter Galling, "Augustine on the Days of Creation: A Look at an Alleged Old-Earth Ally," January 18, 2012, https://answersingenesis.org/days-of-creation/augustine-on-the-days-of-creation.

[12]Rowan Williams, *On Augustine* (New York: Bloomsbury, 2016), vii.

image of a hardened defender of orthodoxy with the realization of a generous, expansive thinker of immense feeling.

RETRIEVING PREMODERN SOURCES FOR MODERN STRUGGLES

But why should we engage Augustine in the first place? The fact that Augustine cared deeply about creation need not entail that we who are interested in creation should care about him. Indeed, is it not an exercise in academic nostalgia to suppose that a figure from the fourth and fifth century can help us address challenges that are largely related to scientific discoveries of the last few centuries?

To be sure, we must recognize Augustine as a man of his times. I, for one, am dismayed at his views on women and sex, amused and perplexed at his speculations about dragons and incubi, and baffled by his theories of time and the soul. Any "retrieval" of Augustine that screens out the oddities of his thought and age will doubtless be artificial.

Nonetheless, I believe that precisely because of their alien feel, premodern theological resources can often serve as a helpful stimulus in contemporary theological work. Quite often, contemporary theology takes its shape in response to essentially *modern* challenges like higher critical theory, or scientific discovery, or secularization. In the case of the doctrine of creation, the table is often set by issues such as how old the world is, whether God created through *de novo* or evolutionary processes, and current environmental and ecological concerns. Because Augustine approached the doctrine of creation prior to the challenges of modernity, his writings can helpfully reframe issues and reorient us to a broader range of concerns. This is one way to locate avenues of thought that might move us beyond the polarization that characterizes much contemporary reflection on the doctrine of creation.

Moreover, we modern Christians must be careful not to assume that we have nothing to learn from the wisdom of the ages. We must guard against the hubris that all knowledge comes through the iPhone. This is what C. S. Lewis called "chronological snobbery."[13] Augustine is a staggeringly deep thinker, and we can benefit from his wisdom, insight, and sincerity. We may even find Augustine to be a helpful corrective against some of the characteristic blind spots of our own age—a kind of palliative or tonic against those eccentricities of modern thought that are invisible to us precisely because they are so close to us.

That is the spirit and motive of this book. It is not a work of historical theology proper; it is a work of *theological retrieval*, which concerns bringing historical theology into the realm of contemporary constructive theology. It attempts to resource current evangelical reflection on the doctrine of creation by retrieving Augustine's vision of creation. To this end, my engagement with Augustine will at times focus on description more than evaluation—not because I endorse everything Augustine says, but because I am seeking to let his voice echo into current controversies.

So picture a long table in a conference room. There are, let us say, three people dialoguing at the table, each representing one of the major creation views today: an evolutionary creationist (or theistic evolutionist, as some prefer to be called) from BioLogos, and old-earth creationist from Reasons to Believe, and a young-earth creationist from Answers in Genesis.[14] Now enters a fourth party: Augustine. The question energizing this book is this: What would

---

[13]C. S. Lewis, *Surprised by Joy* (New York: Harcourt Brace Jovanovich, 1966), 207-8.

[14]I mention these three ministries because they are probably the best-known representatives of the three major current "camps" of creation views among evangelicals. There are different ways to schematize the various creation views in the church today, but this threefold division—young-earth, old-earth, evolutionary creationist/theistic evolutionist—is probably the most recognizable, and it is a wieldy schema that accurately reflects where some of the greatest differences lie. At the same time, it will be helpful to bear in mind lengthier categorizations that draw out greater nuance in the various positions. Some of the different varieties of evolutionary creationism/theistic evolution are drawn out, for instance, in the sixfold categorization of creation views identified by Gerald Rau: (1) naturalistic evolution; (2) nonteleological evolution; (3) planned evolution; (4) directed evolution; (5) old-earth creationism; and (6) young-earth creationism. See the discussion in Matthew Barrett and Ardel B. Caneday, "Introduction," in *Four Views on the Historical Adam*, Counterpoints (Grand Rapids: Zondervan, 2013), 20-25.

Augustine's presence add to the discussion? What would Augustine say that might flavor, deepen, inform, arbitrate, conciliate, or redirect the discussion already being had by people like Francis Collins, Hugh Ross, and Ken Ham?

There are obvious dangers in this approach. Retrieval efforts can easily feel or become contrived. But I have felt that Augustine is too great a teacher, and our need for learning too dire, not to make the attempt. In his recent introduction to Augustine's theology, Matthew Levering suggests that engaging Augustine is not only essential to understand current Roman Catholic and Protestant theology, but also the drift of Western civilization into its "present intellectual impasse," arguing that "Augustine speaks as powerfully today as he did sixteen hundred years ago."[15] If Karl Barth was right to assert that in doing theology "as a member of the church, as belonging to the *congregatio fidelium*, one must not *speak* without having *heard*,"[16] this book is, most basically, an attempt to hear, and help others hear, a voice from within that *congregatio* that must not be ignored.

A presupposition of this effort is that his voice is not always easy to hear. Augustine speaks with a different accent than we do. Understanding him, therefore, requires effort, humility, and patience. But to press the point: Augustine can be valuable *to* us precisely where he is different *from* us. Elsewhere I have described travel to a foreign country as a metaphor for theological retrieval.[17] Becoming immersed in a different culture is a profoundly educational experience, and it provides opportunity for self-perspective—one comes to know the

---

[15]Matthew Levering, *The Theology of Augustine: An Introductory Guide to His Most Important Works* (Grand Rapids: Baker Academic, 2013), xii. More recently, Levering has retrieved Augustine's doctrine of creation to interact with the cosmology of Lawrence Krauss. See Matthew Levering, "Augustine on Creation: An Exercise in Dialectical Retrieval of the Ancients," in *Wisdom and the Renewal of Catholic Theology: Essays in Honor of Matthew L. Lamb*, ed. Thomas P. Harmon and Roger W. Nutt (Eugene, OR: Pickwick, 2016), 49-65.

[16]Karl Barth, *Credo* (New York: Scribner's, 1962), 181; italics his.

[17]See Gavin Ortlund, *Theological Retrieval for Evangelicals: Why We Need Our Past to Have a Future* (Wheaton, IL: Crossway, 2019), 68-73.

peculiarities of one's own culture best in the context of learning about other cultures. No generous-minded person would regard a disinclination to learn from that which is foreign as a sign of progress; rather, it is a sign of narrowness.

Another metaphor for theological retrieval, very much to the same point, is having a conversation. Having a conversation is a qualitatively different experience from private meditation or reflection. Since it involves another person, it contains the possibility for particular kinds of learning that usually are not had by going off alone into the woods and thinking. Of course, private reflection has its place as well. But there is a particular kind of joy and stimulus found in genuinely seeking to understand another person's way of thinking. There is much we can learn from other minds that we would never find searching throughout our own.

Some object to the usage of history in favor of a supposed more neutral historical knowledge. If we approach the past primarily to learn from it, it is felt, we run the risk of distorting it. Of course, this is always a danger. But I would argue that an interest in some kind of "retrieval" or usage of the past is compatible with the best methods of historical research, and to some extent is an inherent part of all historical inquiry. As Marc Bloch famously argued, the best historian is generally one who searches the annals of the past with an alert eye to the present scene as well.[18] R. A. Markus described this in his own engagement with Augustine:

> Any sustained historical enquiry ought to do something to the mind of the historian who undertakes it. It should force him to scrutinize his own assumptions, to question some of his own values, to challenge some of the stock responses of his own age. Prolonged contact with a mind of the stature of Augustine's is inevitably a two-way commerce between the past and the present.[19]

[18]Marc Bloch, *The Historian's Craft*, trans. Peter Putman (New York: Vintage, 1953), 43-47.

[19]R. A. Markus, *Saeculum: History and Society in the Theology of St. Augustine* (Cambridge: Cambridge University Press, 1970), viii.

Nonetheless we must attempt, to the best of our ability, to let Augustine's own voice come through without constraining or reshaping his views to fit contemporary concerns. The polarized nature of creation debates today makes it all the more important to attempt a careful and evenhanded engagement with Augustine, allowing him to speak for himself and taking pains not to squeeze him into the mold of a current ideology or agenda. To the extent that Augustine's contribution, as summarized here, is not wholly aligned with any contemporary "camp," this goal is more likely to have been met.

OUR PATH FORWARD

In chapter 1, we will explore the significance of creation throughout Augustine's theology with a view to how the breadth of Augustine's vision can broaden, shore up, and redirect evangelical engagement with this doctrine. In chapter 2, we will explore the humility Augustine modeled in thinking about creation with a view to its implications for current reflection on the relationship between Scripture and science. The rest of the book then brings Augustine's views into dialogue with particular topics of current dispute—the days of Genesis 1 (chapter 3), animal death (chapter 4), and Adam and Eve (chapter 5).

You will see that I have not attempted to be exhaustive in my engagement with Augustine's doctrine of creation. There are significant strands of his thought that come into the discussion more tangentially here, particularly when they are less involved in some of the contemporary issues on which I am focused. For instance, I have focused less on Augustine's doctrine of original sin, though it impinges on topics addressed in chapter 5. This is obviously an important area of both current dispute and Augustine's legacy. Indeed, for many moderns, original sin is as much associated with Augustine as predestination is with John Calvin and hellfire preaching is with Jonathan Edwards (though such associations, of course, often amount to caricature).

Nonetheless, I have focused less on original sin here for several reasons. First, much has already been written about Augustine and original sin,[20] while other areas of Augustine's doctrine of creation are more commonly neglected. My goal for this project is to focus on some of the issues that have received less treatment but that I believe are no less pressingly relevant to contemporary discussion. Second, in some respects the field of hamartiology (including within its purview the doctrine of original sin) is downstream from creation proper, and thus materially out of scope from many of the topics in view here. Finally, for all the emphasis on Augustine's role in formulating the doctrine of original sin (stemming, as often claimed, from his use of Jerome's mistranslation of Romans 5:12 as "in whom all sinned" instead of "because all sinned"),[21] Augustine's overall anthropology bears many essential continuities with the tradition he inherits, including Tertullian, Cyprian, and Paul himself.[22] As Alan Jacobs suggests, what has made Augustine so notorious is not so much his doctrine of original sin itself but some of the concomitant elements of Augustine's formulation of the doctrine, such as his views on concupiscence and infant damnation: "The whole doctrine of original sin, in Western Christianity anyway, got inextricably tangled with revulsion toward sexuality and images of tormented infants. And there has never been a full and complete disentangling."[23] But many of the pressing areas of concern today relating to original sin are not

---

[20]One of the biggest disputes concerns Augustine's role in the development of this doctrine, and in particular whether Augustine "invented" it or faithfully interpreted and codified the doctrine of original sin in its earlier articulations. For a brief overview of the literature on this point, see Pier Franco Beatrice, *The Transmission of Sin: Augustine and the Pre-Augustinian Sources*, trans. Adam Kamesar (Oxford: Oxford University Press, 2013), 3-8.

[21]For example see Dennis R. Venema and Scot McKnight, *Adam and the Genome: Reading Scripture After Genetic Science* (Grand Rapids: Brazos, 2017), 173.

[22]Alan Jacobs, *Original Sin: A Cultural History* (New York: HarperOne, 2008), 32, pushes against certain scholarly tendencies to juxtapose Paul and Augustine, such that Augustine's view of original sin is abstracted out of its context.

[23]Jacobs, *Original Sin*, 66; cf. also 47.

unique to Augustine, and the basic idea that all are born with a sinful nature because of Adam and Eve's first act of disobedience is hardly an Augustinian "invention."

In addition, despite its importance, I have spent less time addressing the *imago Dei*; Gerald Boersma has recently provided a penetrating study of Augustine's thought in this area.[24] I have also said relatively little about Augustine's theological anthropology,[25] and have made little attempt to defend Augustine from secular critique.[26]

My research for this project has pulled me into three distinct orbits: Augustine's own writings, modern Augustine scholarship, and contemporary literature on the doctrine of creation (with a particular focus on current areas of dispute among evangelicals). Because of the interests driving this book, however, I have leaned more heavily toward the first and third of these fields, engaging and deploying the second (Augustine scholarship) in a more ad hoc manner as it facilitates the conversation I want to nurture between Augustine and the current issues. At any rate, Augustine is such a juggernaut (and the scholarship surrounding him such a thick forest) that it would be difficult to avoid a level of selectiveness even in a strictly historical study as opposed to a retrieval work. In the footnotes, I attempt to at least gesture toward the relevant conversations within Augustine scholarship when I do not pursue them here.

---

[24]Gerald P. Boersma, *Augustine's Early Theology of Image: A Study in the Development of Pro-Nicene Theology*, Oxford Studies in Historical Theology (Oxford: Oxford University Press, 2016). Boersma argues that earlier Latin theologians such as Hilary, Victorinus, and Ambrose associated the image specifically with the Son of God while the participatory ontology that Augustine imbibed from the Neoplatonic tradition enabled him to identify both the Son and humanity as the image. Thus, for Augustine, the image took on greater soteriological significance, particularly in light of Christ's incarnation.

[25]On this topic see Matthew Drever, "Image, Identity, and Embodiment: Augustine's Interpretation of the Human Person in Genesis 1–2," in *Genesis and Christian Theology*, ed. Nathan MacDonald, Mark W. Elliott, and Grant Macaskill (Grand Rapids: Eerdmans, 2012), 117-28; G. R. Evans, *Augustine on Evil* (Cambridge: Cambridge University Press, 1982), 42-47.

[26]For a sympathetic engagement with Augustine in relation to criticisms in the traditions of Kant, Hegel, Freud, Nietzsche, and Heidegger, see Paul Rigby, *The Theology of Augustine's Confessions* (Cambridge: Cambridge University Press, 2015).

In terms of Augustine's writings, I have drawn from a wide array of published works as well as sermons and letters, with a more systematic emphasis on the five works I regard as most significant for his doctrine of creation:

- *On Genesis: A Refutation of the Manichaeans* (*De Genesi contra Manichaeos*), written around 388–389
- *The Unfinished Literal Commentary on Genesis* (*De Genesi ad litteram liber unus imperfectus*), written around 393–395
- *Confessions* (*Confessiones*), written around 397–401
- *The Literal Meaning of Genesis* (*De Genesi ad litteram*), written around 401–416
- *The City of God* (*De civitate Dei*), written around 413–426

My engagement with the contemporary literature surfaces throughout, and is particularly the burden of chapter 5.

My hope is that this book will be a resource for the study of the doctrine of creation, will contribute to Augustine scholarship, and will encourage humility, carefulness, and conviction in the church today.

To whet your appetite for moving forward, hear again Augustine's testimony, this time as he recounts it himself in his famous *Confessions*:

> So I came to Milan, to the bishop and devout servant of God, Ambrose. . . . That man of God received me as a father, and as bishop welcomed my coming. I came to love him, not at first as a teacher of the truth, which I had utterly despaired of finding in Your Church, but for his kindness towards me. I attended carefully when he preached to the people. . . .
>
> I began to see that the Catholic faith, for which I had thought nothing could be said in the face of the Manichaean objections, could be maintained on reasonable grounds: this especially after I had heard explained figuratively several passages of the Old Testament which had been a cause of death for me when taken literally.

Many passages of these books were expounded in a spiritual sense and I came to blame my own hopeless folly in believing that the law and the prophets could not stand against those who hated and mocked at them.[27]

[27]*Confessiones* 5.13-14 (CSEL 33, 110-11); translation from F. J. Sheed in Augustine, *Confessions*, ed. Michael P. Foley, 2nd ed. (Indianapolis, IN: Hackett, 2006), 90-91.

CHAPTER ONE

WHAT WE FORGET ABOUT CREATION

How Augustine Can Broaden our Horizon of Concerns

Let me hear and understand the meaning of the words:
In the Beginning you made heaven and earth. Moses wrote these words. . . . If he were here, I would lay hold of him and in your name I would beg and beseech him to explain those words to me. I would be all ears to catch the sounds that fell from his lips.

Confessiones 11.3

Creation is a frequently underdeveloped, atrophied doctrine. John Webster has spoken of the "cramping effects" that modernity imposes on theology, identifying two particular loci where the damage can be seen: the Trinity and the doctrine of creation.[1] Often Christians treat the doctrine of creation as a kind of prolegomenon to theology rather than a theological topic in its own right. Creation is important, it is thought, primarily insofar as it sets the stage for the weightier matters of theology—those issues involved in the doctrine of redemption.

[1]John Webster, "Theologies of Retrieval," in *The Oxford Handbook to Systematic Theology*, ed. John Webster, Kathryn Tanner, and Iain Torrance (Oxford: Oxford University Press, 2008), 594-95.

When we do engage the theology of creation more directly, interest is often narrowly focused on questions springing from science-faith dialogue: What is the nature of the days in Genesis 1? Are the Adam and Eve of Genesis 2–3 historical figures? Was there a historical fall, and how do we understand this event in relation to the claims of evolutionary science?[2]

These are obviously vital questions. However, if we engage Genesis 1–3 as more than a mere preamble or preface to the biblical story, we will find that the material contribution of these chapters to Christian theology is far from exhausted by such concerns. This portion of Scripture offers a holistic framework for how to live as God's creatures in God's world, helping us integrate every aspect of our existence—relationships, work, art, laughter, music, play—under our calling as God's image-bearers.

In the church, we have often emphasized the Christian life without reference to life as a human being.[3] But the categories of sin and salvation are only comprehensible in light of the prior category of creation—the assertion "I am a sinner" is a further specification from the assertion "I am a creature." Furthermore, if redemption involves

---

[2]More than two decades ago, Colin Gunton observed that "it is not too much of an exaggeration to say that in the modern world the doctrine of creation has in many places given way to discussion of the relation between science and religion" ("Introduction," in *The Doctrine of Creation: Essays in Dogmatics, History, and Philosophy*, ed. Colin Gunton [London: T&T Clark International, 1997], 1). Similar interests are reflected in much contemporary evangelical theology. Consider, for instance, the following four books, all published in 2017, all significantly engaged by evangelical theologians and leaders, and all devoted to exploring questions in this vein of concerns: J. B. Stump, ed., *Four Views on Creation, Evolution, and Intelligent Design*, Counterpoints (Grand Rapids: Zondervan, 2017); Kenneth Keathley, J. B. Stump, and Joe Aguirre, eds., *Old-Earth or Evolutionary Creation?: Discussing Origins with Reasons to Believe and BioLogos*, BioLogos Books on Science and Christianity (Downers Grove, IL: IVP Academic, 2017); J. P. Moreland, Stephen C. Meyer, et al., eds., *Theistic Evolution: A Scientific, Philosophical, and Theological Critique* (Wheaton, IL: Crossway, 2017); Theodore J. Cabal and Peter J. Rasor II, *Controversy of the Ages: Why Christians Should Not Divide Over the Age of the Earth* (Wooster, OH: Weaver, 2017). The issues explored in these volumes are obviously important, and these books make a helpful contribution to these discussions. Nonetheless, it is unfortunate that we do not see *more* books on *other* aspects of the doctrine of creation. Such evangelical intramural debates hardly exhaust the dogmatic significance of the topic.

[3]One consequence may be our frequently underdeveloped doctrines of vocation, the arts, and what it means that we are embodied creatures.

not a repudiation of our original creaturely mandate but rather a re-orientation toward it (e.g., Col 3:10; Eph 4:24), then the doctrine of creation not only precedes and undergirds the doctrine of redemption, but informs it. We are not just saved *from* something (sin), but saved *to* something (imaging God).

In this chapter, we suggest that retrieving Augustine's doctrine of creation is one way to broaden our horizon of concerns in this area. Now, there are many aspects of the doctrine of creation that could be happily welcomed into a more prominent position within evangelical consciousness and dialogue—say, providence, or angelology, or the contingency of creation, or the goodness of creation, or ecological concern, or trinitarian agency in creation. Some of these topics are engaged more by evangelical academics than evangelical laity (perhaps, e.g., contingency); some tend to be altogether underworked (perhaps, e.g., angels); in all of them, arguably, Augustine could be useful.

But here we will focus on one issue: the ontological shape of Augustine's doctrine of creation and its implication for human happiness. Augustine's treatment of creation emphasized a thick distinction between God and his creation (what we will call *divine priority*), with a consequent radical dependence of the world upon God (what we will call *creaturely contingency*), and there is a sense in which this ontological framework drives everything else in Augustine's theology. Engaging this aspect of Augustine's thought, even where we do not finally agree with him, may helpfully draw attention to the pervasive significance of the doctrine of creation throughout Christian theology. Here we will emphasize in particular how creation was, for Augustine, the clue to unraveling the deepest longings of the human heart.

We start in this way so that we will not immediately demand that Augustine give us answers, but first give him an opportunity to reshape and reformulate our questions, perhaps pulling us into new directions and broadening our horizon of interest. As an entry point, we can make the issue more pressing by drawing attention to the often-neglected significance of the doctrine of creation in Augustine's most famous work, the *Confessions*.

CONFESSIONS OF HEART AND UNIVERSE

Interpreters of Augustine's *Confessions* have often puzzled over the book's ending. After nine chapters of intensely personal autobiography, why does Augustine then conclude with more abstract accounts of memory (book 10) and creation (books 11–13)? Or, more typically, how does Augustine move from himself (books 1–10) to Genesis (books 11–13)? The transition feels somewhat abrupt, both in tone and content. Indeed, according to Jared Ortiz, the scholarly "consensus" over the last century or so is that "the *Confessions* does not have a singular meaning and that it does not hold together."[4] Thus, John J. O'Meara claims that it is "a commonplace of Augustinian scholarship to say that Augustine was not able to plan a book."[5] Not surprisingly, this critical view of the *Confessions*' literary and structural integrity often results in more scattered and piecemeal engagements with its content.[6] A book so badly written, after all, need not be read *too* carefully.

But there are reasons to doubt that Augustine would have perceived such a tension between the abstract and emotional qualities of his book. Marjorie O'Rourke Boyle has drawn attention to Cicero's influence on Augustine's style of rhetoric and argument, suggesting that the *Confessions* is "composed quite classically according to the ordinary Ciceronian rules for the invention of argument which

[4]Jared Ortiz, *"You Made Us for Yourself": Creation in St. Augustine's Confessions* (Minneapolis: Fortress Press, 2016), xvii. Ortiz traces this claim from the early criticism of Henri Marrou, who famously charged that *Augustin compose mal* ("Augustine writes badly"), to modern Augustine scholars like James O'Donnell, who critiques efforts to find a "key" that can unlock the structure of the *Confessions* and reveal an underlying unity. He also notes P. L. Landsberg's influential attempt to identify the unity of the work in its method reflected in its title: *confessio*. See Ortiz, *"You Made Us for Yourself,"* xvii-xxi.

[5]John J. O'Meara, *Young Augustine: The Growth of St. Augustine's Mind Up to His Conversion* (London: Longmans, Green, 1954), 44.

[6]As noted by Ortiz, *"You Made Us for Yourself,"* xxi.

Augustine habitually practiced as rhetor, then preacher."[7] With respect to the book's development of thought, Robin Lane Fox argues that though these final chapters seem to be on a higher plane, they are not "additions to an 'autobiographical' work," but rather "the culmination of the entire work," since Augustine's meditation on time and creation in books 10–13 represents the fulfillment of his longing for worship, the great pursuit of books 1–9.[8]

Similarly, Ortiz suggests that Augustine's doctrine of creation is actually the key to the whole book by situating Augustine's story in relation to his larger vision of reality. In ancient thought, with the exception of Christians and Platonists, God tended to be conceived as one part of the world, rather than transcendent over it.[9] The Christian notion of creation *ex nihilo* signaled a fundamentally different structure to reality: it meant that "for anything to be, it has to be drawn back to God so it can share in his being in some way. Only by becoming like God can things be."[10] By its very nature, an *ex nihilo*, contingent creation can only be and become what it is through a continual turning to the One who made it and sustains it. This broader ontology helps illumine why Augustine had to "confess" his personal salvation and his view of the universe together. As Henry Chadwick puts it:

> Augustine understood his own story as a microcosm of the entire story of creation, the fall into the abyss of chaos and formlessness, the "conversion" of the creaturely order to the love of God as it experiences griping pains of homesickness. What the first nine books [of the *Confessions*] illustrate in his personal exploration of the experience of the prodigal son is given its cosmic dimension in the concluding parts of the work.[11]

---

[7]Marjorie O'Rourke Boyle, "The Prudential Augustine: The Virtuous Structure and Sense of his Confessions," *Recherches Augustiniennes* 22 (1987): 131.

[8]Robin Lane Fox, *Augustine: Conversions to Confessions* (New York: Basic Books, 2015), 544-45.

[9]Ortiz, *"You Made Us for Yourself,"* xxiv.

[10]Ortiz, *"You Made Us for Yourself,"* 230.

[11]Henry Chadwick, *Augustine: A Very Short Introduction* (Oxford: Oxford University Press, 1986), 70.

In this way of thinking, the *Confessions* has a profoundly coherent internal unity, from the initial famous declaration of its opening paragraph ("You have made us for yourself and our hearts are restless until they rest in you"[12]) to the concluding focus of book 13 on divine rest in Genesis 1 as the end of creaturely restlessness ("When our work in this life is done, we too shall rest in you in the Sabbath of eternal life"[13]).

But how do these personal and cosmic dimensions relate more precisely? What was the vision of creation that enabled Augustine to correlate his own soul's "restlessness" with the creaturely tilt of the entire created order, from the plainest pebble to the brightest angel?

In what follows we will briefly trace this motif of creaturely happiness ("rest"), particularly as seen in his famous *Confessions*. We will then situate it in relation to Augustine's broader ontological framework for creation, focusing on five principles: divine priority, creaturely contingency, trinitarian agency in creation, sin as privation, and redemption as deification. Finally, we derive three specific conclusions for how Augustine's views in this area might help broaden evangelical reflection on the doctrine of creation today.

CREATURELY "RESTLESSNESS" THROUGHOUT THE *CONFESSIONS*

The motif of creaturely "rest" is not limited to Augustine's *Confessions*. For instance, in his finished commentary of Genesis, Augustine describes creatures as good yet imperfect, in need of sharing in God's "quiet rest." He insists that the perfection of each created thing occurs not in the whole of which it is a part, but rather in him from whom it derives its being.[14] He describes each created thing "finally coming to rest" in God as the attainment of "the goal of its own momentum." The "momentum" he has in view here is generated

[12] *Confessiones* 1.1 (CSEL 33, 1); my translation.

[13] *Confessiones* 13.36 (CSEL 33, 387). Cf. his similar treatment of divine rest in *De civitate Dei* 11.8 (CSEL 40.1, 521-22) and 11.31 (CSEL 40.1, 559-60).

[14] *De Genesi ad litteram* 4.18.34 (CSEL 28:1, 117).

by creatureliness—the inherent tilt of all creatures toward God. Thus, Augustine continues:

> The whole universe of creation . . . has one terminus in its own nature, another in the goal which it has in God. . . . It can come to no stable and properly established rest, except in the quiet rest of the one who does not have to make any effort to get anything beyond himself to find rest in it. And for this reason, while God abides in himself, he swings everything whatever that comes from him back to himself, like a boomerang, so that every creature might find in him the final terminus and goal for its nature, not to be what he is, but to find in him the place of rest in which to preserve what by nature it is in itself.[15]

Here Augustine distinguishes between two different termini or goals of creatures: one in their own nature, and one that final state of entering God's rest. He emphasizes the incompleteness of creatures' own terminus, claiming they lack any "stable and properly established rest," and contrasts this with God's self-sufficiency as the one who has rest in himself, the one who does not need anything beyond himself in order to find rest. Moreover, strikingly, Augustine depicts God as continually at work in relation to this ontological divide, swinging everything he has created back to himself to find rest in himself. Although the translator, Edmund Hill, has added the boomerang imagery here, it captures something of Augustine's meaning: God creates imperfect creatures with an inherent need for him, and subsequently "swings" them back to himself. Thus, creation must return to its source in order to find itself. Every creature must return to its Creator, like a boomerang, to preserve its own nature.

Now, when did this rest begin, and when does it conclude? Augustine does regard creation to have begun this activity of sharing the Creator's rest after the evening of the sixth day, but he also holds that it will continue to develop until it finds a secure and final rest in him.

[15]*De Genesi ad litteram* 4.18.34 (CSEL 28:1, 117).

In this final state, all of creation will abide forever, since anything that has existence only has it through participation in God: "Since what the whole created universe is going to be, whatever mutations it has gone through, will certainly not be nothing, the whole created universe will, for that reason, always abide in its creator."[16]

Consider an analogy for this way of thinking about creation: Suppose an artist is constructing a piece of pottery. He completes the work, but it has not yet gone through the firing and glazing stages, which transform it from being a soft, breakable artifact into durable, usable ceramic. In one sense, it is complete; in another sense, it is not. It has been fully shaped, but it has not yet come into its proper goal. Augustine thinks the entire universe is like this: incomplete in a crucial sense, tilted forward toward its final goal. Only the artisan can complete it; the pottery cannot fire and glaze itself. So our world has no "rest" in itself.

This language of "restlessness" calls to mind, of course, Augustine's famous prayer at the start of the *Confessions*, "You have made us for yourself and our hearts are restless until they rest in you."[17] For all its quotability, this statement is one way to sum up Augustine's entire framework for creation, in which God's work of creation ("you have made us for yourself") imbues an inherent need ("our hearts are restless") for a further union with God ("until they rest in you"). It may be useful to reflect a bit on how this justifiably famous quotation coheres with themes that run throughout the rest of the *Confessions*.

To begin with, it is worth noting that this statement comes at the very beginning of the *Confessions*, closely following Augustine's reference to human death as a "sign" and "reminder" of human sin.[18] In context, Augustine appears to be attempting to diagnose the human heart in its current, postlapsarian setting. Yet Augustine also sees the

[16]*De Genesi ad litteram* 4.18.35 (CSEL 28:1, 118).
[17]*Confessiones* 1.1 (CSEL 33, 1); my translation.
[18]*Confessiones* 1.1 (CSEL 33, 1).

impulse in the human heart to worship as a consequence of our creatureliness, not our fallenness: "Man is one of your creatures, and his instinct is to praise you;"[19] "Since he is a part of your creation, he wishes to praise you."[20] In the subsequent paragraphs of the opening of the *Confessions*, Augustine somewhat problematizes this creaturely instinct toward praise, wondering aloud whether he should begin with supplication for praise or with praise itself. This dilemma is generated by God's unknowability: "If (a man) does not know you, how can he pray to you?"[21] It is also the implication of creaturely finitude: "What place is there in me in which my God can come? Where can God come into me—God, who made heaven and earth?"[22] To be sure, Augustine knows that God is within him; if he were not, Augustine could not exist: "I should be null and void and could not exist at all, if you, my God, were not in me."[23] Yet God is, at the same time, infinitely distant—he is both "the most hidden and most present."[24]

Augustine stacks up paradoxes implicit in the God-world relation to emphasize the dilemma of his situation: God is the essence of creaturely happiness, and simultaneously beyond creaturely capacity. We were made for God, but cannot hold him. He alone can fill us, but we cannot contain him. Thus, as Augustine sees it, creatureliness has an inherent *unsettledness* to it: the very thing for which we have been created is beyond our grasp, and nothing else can fill its void. Moreover, this unsettledness is equally characteristic of every particular creature as it is for the entire creation. Hence Augustine will revert back and forth throughout these passages between the "restlessness" of his own soul and that of all heaven and earth.

This paradox of God's simultaneous necessity and impossibility is the tension that drives the *Confessions*, and ultimately stipulates the

[19]*Confessiones* 1.1 (CSEL 33, 1).
[20]*Confessiones* 1.1 (CSEL 33, 1).
[21]*Confessiones* 1.1 (CSEL 33, 1).
[22]*Confessiones* 1.2 (CSEL 33, 2); my translation.
[23]*Confessiones* 1.2 (CSEL 33, 2).
[24]*Confessiones* 1.4 (CSEL 33, 3); my translation.

prayerful method Augustine employs throughout: "I shall look for you, Lord, by praying to you and as I pray I shall believe in you."[25] The famous prayerful genre of the *Confessions* should therefore not be considered a mere literary consideration for Augustine, but rather a *theologically* informed choice occasioned by this challenge of God's simultaneous importance and distance. Prayer is Augustine's tool to look for the invisible, yet ever-near, God.

Now, those who see the *Confessions* as little more than an autobiography will doubtless be perplexed by its later chapters. But if we consider Augustine's own personal testimony in books 1–9 as merely one step in the larger effort of his pursuit of creaturely praise and "rest," what might we find in books 10–13 that coheres with this theme? After his account of his conversion in book 8, Augustine passes through his subsequent stay at Cassiacum and his baptism in Milan relatively rapidly, reflecting much on his mother, Monica, in book 9, then proceeding to his discussion of memory in book 10. At this point the strictly narrative parts of the *Confessions* end, and Augustine transitions from confessing his past sins to confessing who he is in the present:

- "I know what profit I gain by confessing my past, and this I have declared. But many people . . . wish to hear what I am now, at this moment, as I set down my confessions."[26]
- "I go on to confess, not what I was, but what I am."[27]
- "I shall therefore confess both what I know of myself and what I do not know."[28]

The reason that Augustine then takes up a lengthy discussion of the topic of memory throughout the subsequent parts of book 10 is that the memory is a faculty of the soul, and it is through the soul that God must be approached: "If I am to reach him, it must be through my

[25]*Confessiones* 1.1 (CSEL 33, 2).
[26]*Confessiones* 10.3 (CSEL 33, 228).
[27]*Confessiones* 10.4 (CSEL 33, 230).
[28]*Confessiones* 10.5 (CSEL 33, 231).

soul."[29] This rationale for exploring his memory, together with the fervent expressions of love for God and confession of his present sins throughout book 10 (especially the later parts), suggest that in book 10 Augustine has not taken leave from the concerns driving the earlier portions of the book. Rather, Augustine searches within himself, in his memory and soul, precisely to look for God. As he stipulates, "I shall go beyond this force that is in me, this force which we call memory, so that I may come to you, my Sweetness and my Light."[30] Thus, the structural transition occurring in book 10 indicates that Augustine is now looking within himself in the present rather than looking back on his life—but nonetheless, both movements share the same overall goal. It is the soul's search for rest in God that motivates Augustine's interest in his memory, five senses, and present sins, no less than his interest in looking back at his conversion.

Books 11–13, then, dealing with the Genesis creation account, represent the culmination of these interests. As Augustine opens 11.1, he looks back on what he has achieved thus far with a sense of completion: "To the best of my power and the best of my will I have laid this long account before you, because you first willed that I should confess to you, O Lord my God."[31] Yet Augustine is not done. After a long petition for divine help in meditating on God's law (11.2), he initiates a new question at the start of 11.3:

> Let me hear and understand the meaning of the words: In the Beginning you made heaven and earth. Moses wrote these words. He wrote them and passed on into your presence, leaving this world where you spoke to him. He is no longer here and I cannot see him. But if he were here, I would lay hold of him and in your name I would beg and beseech him to explain those words to me. I would be all ears to catch the sounds that fell from his lips.[32]

---

[29]*Confessiones* 10.7 (CSEL 33, 234).
[30]*Confessiones* 10.17 (CSEL 33, 246).
[31]*Confessiones* 11.1 (CSEL 33, 280-81).
[32]*Confessiones* 11.3 (CSEL 33, 283).

It is striking that here, in this final segment of the *Confessions*, Augustine turns his gaze on (of all things) Genesis 1:1. Careful attention to how he proceeds, however, suggests that this is not a turn away from the book's deeper themes, but Augustine's pathway to enter more deeply into them. For instance, throughout book 11 Augustine pursues the nature of time, particularly in relation to Manichaean objections such as the question of what God was doing before creation. Yet time is interesting to him not as an abstract problem, but in relation to his desire to enter into God's eternal rest. This helps explain why his discussion throughout book 11, even while pursuing complex philosophical questions, retains the same emotional urgency that characterizes the rest of the *Confessions*. Thus we hear Augustine declaring that his mind is burning with curiosity to understand the nature of time,[33] bewailing the sorry state of his ignorance,[34] and calling for resoluteness of soul to pursue the answer.[35] Similarly, when he engages Genesis 1:1-2 in book 12, he describes it as a passage that has "set my heart throbbing."[36]

Why is the nature of time such a personally affecting doctrine for Augustine? For Augustine, creaturely happiness consists of entering into God's eternal rest. Temporality entails mutability, and mutability entails continually falling away from the immutable God who is the source of all goodness. The supreme goal of every creature is therefore to enter into God's eternal, unchanging rest, and thus share (in a sense) in God's immutability. This is what the spiritual/intellectual creation—i.e., the realm of angels, which Augustine terms the "heaven of heaven"[37]—already possesses in its constant clinging to God.[38] It is this

---

[33]*Confessiones* 11.22 (CSEL 33, 299).
[34]*Confessiones* 11.25 (CSEL 33, 303).
[35]*Confessiones* 11.27 (CSEL 33, 304).
[36]*Confessiones* 12.1 (CSEL 33, 310).
[37]Augustine distinguishes the "corporeal heaven" (that is, the stars and the physical space they occupy) from what he calls, drawing upon the Vulgate rendering of Psalm 115:16, the *caelum caeli*, the "heaven of heaven" or the "incorporeal heaven of the corporeal heaven"—that is, the dwelling place of God and angels. Cf. *De Genesi ad litteram* 1.17.32 (CSEL 28:1, 24).
[38]*Confessiones* 12.11 (CSEL 33, 317).

participation in God's immutability that is the ground of angelic happiness: "How happy must this creature be, if such it is, constantly intent upon your beatitude, forever possessed by you, forever bathed in your light!"[39] Augustine insists that the spiritual creation is not coeternal with God and possesses a derived, not an inherent, immutability. Thus, although it does not change, "Mutability is inherent in it, and it would grow dark and cold unless, by clinging to you with all the strength of its love, it drew warmth and light from you like a noon that never wanes."[40] For Augustine, the brightest seraphim in the Heaven of Heavens and the lowest worm in the earth are, for all their differences, equally *contingent* creatures: they grow dark and cold, and ultimately veer off into nothingness, if they do not continually cling to God.

It is this interest in the creaturely happiness that comes by sharing in divine rest that drives books 11–13 of the *Confessions.* Augustine describes two distinct kinds of temporality available to creatures: the spiritual creation, which is "mutable but without mutation" because "it is constant in its enjoyment of your eternity and absolute immutability," and the lower creation, which Augustine associates with the "formlessness" from Genesis 1:2 because it is mutable.[41] Augustine, as a man, belongs to this lower, physical order marked by mutability, but what his soul longs for is to pass over into God's immutable rest:

> Thou, O Lord, my eternal Father, art my only solace; but I am divided up in time, whose order I do not know, and my thoughts and the deepest places of my soul are torn with every kind of tumult until the day when I shall be purified and melted in the fire of Thy love and wholly joined to Thee.[42]

Here Augustine envisions ultimate creaturely happiness as a kind of fusion with God, passing from a mutable state into sharing in divine

[39]*Confessiones* 12.11 (CSEL 33, 317-18).
[40]*Confessiones* 12.15 (CSEL 33, 323).
[41]*Confessiones* 12.12 (CSEL 33, 319).
[42]*Confessiones* 11.29 (CSEL 33, 308); translation from F. J. Sheed, *Confessions*, 255.

immutability. At an important section of *The City of God*, at the very beginning of his development of the doctrine of two cities, Augustine describes a similar goal: "It is a great and very rare thing for a man, after he has contemplated the whole creation, corporeal and incorporeal, and has discerned its mutability, to pass beyond it, and, by the continued soaring of his mind, to attain to the unchangeable substance of God."[43]

In Augustine's vision of creation, it is not only his soul but all of creation that longs to enter into God's eternal, immutable happiness. This is why book 13 of the *Confessions* climaxes in an allegorical reading of Genesis 1, with God's rest on the seventh day of creation representing the eternal joy of heaven into which creatures may enter: "You rested on the seventh day. And in your Book we read this as a presage that when our work in this life is done, we too shall rest in you in the Sabbath of eternal life."[44] As we have noted, this suggests the incompleteness of physical creation—Augustine regards it as good, but it stands in need of a further act to enter into God's immutability with the spiritual creation and attain its goal. All creaturely existence is oriented toward the eternal divine Sabbath.

This is the great arc that energizes the *Confessions*: from the "we are restless until we find our rest in you" at its beginning to the "we too shall rest in you in the Sabbath of eternal life" of its conclusion. This is why the doctrine of creation is so foundational for Augustine's theology, and why Genesis 1 is a passage that sets his heart throbbing.

DIVINE PRIORITY

What was the broader view of creation in which this notion of creaturely restlessness functioned for Augustine? In the first place, it was a derivation of his conviction of God's ontological priority. Divine

---

[43]*De civitate Dei* 11.2 (CSEL 40.1, 512).

[44]*Confessiones* 13.36 (CSEL 33, 387); cf. his similar treatment of divine rest in *De civitate Dei* 11.8 (CSEL 40.1, 521-22) and 11.31 (CSEL 40.1, 559-60).

ontological priority simply means that God's being is qualitatively prior to all else, which entails a thick ontological distinction between God and creation. Thus reality breaks down, at the most fundamental level, as twofold rather than one, legion, or indeterminate. There are essentially two kinds of being: God and everything else.[45] Moreover, these two kinds of being are utterly distinct. The distance between them is infinite.[46] Whatever the differences are between two creatures, the ontological gulf between both of them and God is infinitely greater. The most significant fact about anything, therefore, is whether or not it is God.

As basic as this point may seem, it is decisive for Augustine's theological outlook. First and foremost, divine priority was bound up with Augustine's articulation of creation *ex nihilo*.[47] This doctrine was not a constructive tool that Augustine invented in the abstract. Rather, *ex nihilo* was developed as an alternative to rival views of creation offered in, for instance, the Manichaean, Parmenidean, and Platonic systems.[48] For instance, in Plato's *Timaeus*, the demiurge shapes the world from matter existing previously in some inchoate form and, constrained to avoid envy, must build the best possible world.[49] In

[45] I had thought of using the term *ontological dualism* instead, but I worry that this could be associated with various other kinds of dualism that are popular objects of critique and thus cause confusion.

[46] Throughout the Christian tradition divine priority is often depicted in spatial categories—for example, John of Damascus famously stipulated that "all things are distant from God not by place, but by nature." Augustine also speaks in this way, and intriguingly states that the angelic spiritual creation is likewise "above every kind of body, not by degrees of space, but by the sublimity of its nature" (*De Genesi ad litteram* 1.17.32 [CSEL 28:1, 24]).

[47] N. Joseph Torchia, *Creatio Ex Nihilo and the Theology of Augustine: The Anti-Manichaean Polemic and Beyond*, American University Studies 205 (New York: Peter Lang, 1999), ix, argues that in a very real sense creation *ex nihilo* "constitutes a crucial, if not the pivotal, element in his theological deliberations on a wide variety of topics." Later he claims that creation *ex nihilo* "established a framework in which the fundamental aspects of Augustine's theology developed" (231).

[48] Augustine was hugely influenced by Platonic thought, and it leaves a stamp everywhere in his writings. By contrast, as noted by Portalié in *A Guide to the Thought of Saint Augustine*, 95, he only cited Aristotle three times, and is generally critical of the Stoics and Epicureans whenever he cites them.

[49] For a helpful discussion of *Timaeus*, which Augustine engaged in *The City of God*, see Torchia, *Creatio Ex Nihilo*, 22-30.

contrast, Augustine insisted that God creates the world *ex nihilo*, from the freedom and overflow of love rather than the avoidance of envy, and that God makes it good but not perfect.[50] Similarly, Augustine articulated creation *ex nihilo* in contrast to the speculations of Aristotle and Parmenides about what God was doing before he created the world, and he affirmed God's freedom to choose to create over and against the influence of Plotinus's theory of emanation.[51]

Above all, Augustine drew on creation *ex nihilo* to emphasize God's transcendence over the world and to combat various Manichaean errors. In his allegorical commentary on Genesis, Augustine repeatedly emphasized the necessity of creation *ex nihilo* for divine omnipotence. Contrasting God's creative work with the potter who works with preexistent clay and the carpenter who builds with preexistent wood, he claimed:

> The carpenter doesn't make wood, but makes something out of wood; and so with all other such craftsmen. But Almighty God did not need the help of any kind of thing at all which he himself had not made, in order to carry out what he wished. If, you see, for making the things he wished, he was being assisted by some actual thing which he had not made himself, then he was not almighty; and to think that is sacrilege.[52]

There is thus a broader body of discussion concerning the God-world relationship that served as a backdrop to Augustine's emphasis on creation *ex nihilo*—just as today the doctrine of creation *ex nihilo* is often a linchpin upon which broader philosophical issues turn.[53]

---

[50]See a discussion of these three points of contrast in William E. Mann, "Augustine on Evil and Original Sin," in *The Cambridge Companion to Augustine*, ed. David Vincent Meconi and Eleonore Stump, 2nd ed. (Cambridge: Cambridge University Press, 2014), 99-102.

[51]For a helpful discussion see Simo Knuuttila, "Time and Creation in Augustine," in *The Cambridge Companion to Augustine*, 83-86, and Torchia, *Creatio Ex Nihilo*, 36.

[52]*De Genesi contra Manichaeos* 1.6.10 (CSEL 91, 76-77).

[53]Cf. David B. Burrell, "*Creatio Ex Nihilo* Recovered," *Modern Theology* 29 (2013): 18, in expounding the Thomist view: "The relation between Creator and creatures . . . is unlike any causal relation we know since God's causation in creating produces [in God] no change or motion or succession in time; rather it is instantaneous, like the burning of fire or the sun lighting up the atmosphere."

(Of course, Augustine's relationship to these alternative systems was not wholly negative; he borrowed certain elements, particularly from Platonic views.[54])

Nonetheless, however it may have been polemically occasioned, creation *ex nihilo* (and the ontological priority of God it entailed) came to function constructively for Augustine's own theology. In the first place, Augustine's emphasis on God as the fount of all being entailed that creation is itself an act of divine generosity. No need compelled God to create, nor did creation fill a "gap" that previously existed. The motive of creation is sheer divine love. This point comes up again and again for Augustine—for example, in his exposition of the creed in the *Enchiridion*, where Augustine emphasizes that the cause of creation is the overflow of divine goodness;[55] or at the end of *De vera religione*, where Augustine affirms that nothing would have been created if God were not "so good that he is not jealous of any nature's being able to derive its goodness from him."[56] Or as Augustine puts it in the *Confessions*, "You created, not because you had need, but out of the abundance (*ex plenitudine*) of your own goodness. You molded your creation and gave it form, but not because you would find your happiness increased by it."[57] Existence is therefore, for anything other than God, wholly superfluous, a kind of mysterious gift. As Vladimir Lossky puts it, "We might say that by creation

[54]For instance, his conception of divine ideas functions something like Platonic forms, such that he can declare that God's "ideas are certain principal forms or stable and unchangeable reasons of things, which are not themselves formed—and so are eternal and always holding themselves in the same mode—which are contained in the divine intelligence" (as quoted in Ortiz, *"You Made Us for Yourself,"* 17). Nonetheless, there are differences between divine ideas and Plato's forms proper—for instance, as Rowan Williams puts it, the "forms" Augustine considers are God's own thoughts rather than "independent models for the divine artisan to work from" (Rowan Williams, "Creation," in *Augustine Through the Ages: An Encyclopedia*, ed. Allan D. Fitzgerald [Grand Rapids: Eerdmans, 1999], 253). For a helpful discussion of the role of divine ideas in creation, see Levering, *Engaging the Doctrine of Creation*, 29-71.

[55]*Enchiridion* 3.9, in *On Christian Belief*, ed. Boniface Ramsey, trans. Bruce Harbert, The Works of Saint Augustine: A Translation for the 21st Century (Hyde Park, NY: New City Press, 2005), 277-78.

[56]*De vera religione* 55.113 (CSEL 77, 80).

[57]*Confessiones* 13.4 (CSEL 33, 348).

ex nihilo God 'makes room' for something that is wholly outside of Himself; that, indeed, He sets up the 'outside' or nothingness alongside of His plenitude."[58]

Another consequence of God's ontological priority for Augustine's theology concerns the Creator-creation relationship. Specifically, divine priority grounded Augustine's paradoxical claim that God is both infinitely near to, and infinitely far from, every creature. For Augustine, the ontological gulf between God and creation does not entail that creation is cut off from the knowledge of God; just the opposite. Following the thought of the apostle Paul in Romans 1:20, Augustine understood creation's contingent status as an act of revelation. As he puts it in the *Confessions*, "I asked the whole mass of the universe about my God, and it replied, 'I am not God. God is he who made me.'"[59] And later: "Earth and the heavens also proclaim that they did not create themselves. 'We exist,' they tell us, 'because we were made. And this is proof that we did not make ourselves. For to make ourselves, we should have had to exist before our existence began.'"[60] Thus, God's fundamental otherness does not enshroud him in darkness, but instead is the principle by which he is undeniably known. As Ortiz puts it, "The ontological distinction between God and the world is a kind of light or epiphany which illumines all things. Creation is a revelation."[61]

Similarly, for Augustine, the ontological gulf that separates God and creatures is paradoxically that which allows him to be "with" his creatures in the preexistent state of his perfect foreknowledge. Augustine argues that all things exist in the mind of God before they are created, and are "better" and "truer" there, since his knowledge of them is eternal and unchangeable. They are, in a sense, with God in

[58]As quoted in Matthew Levering, *Engaging the Doctrine of Creation: Cosmos, Creatures, and the Wise and Good Creator* (Grand Rapids: Baker Academic, 2017), 46.
[59]*Confessiones* 10.6 (CSEL 33, 233).
[60]*Confessiones* 11.4 (CSEL 33, 284).
[61]Ortiz, *"You Made Us for Yourself,"* xxiv-xxv.

this preexistent state, and thus "alive."[62] Augustine develops this thought in an interesting progression. Quoting the "I am who I am" of Exodus 3:14, Augustine asserts that "the manner in which [God] is differs totally from that in which these things are that have been made."[63] He then reasons:

> So without bringing into existence yet any of the things which he made, [God] has all things primordially in himself in the same manner as he is. After all, he would not make them unless he knew them before he made them; nor would he know them unless he saw them; nor would he see them unless he had them with him; and he would not have with him things that had not yet been made except in the manner in which he himself is not made.[64]

This intriguing chain of reasoning moves from God's creating to God's knowledge, from God's knowledge to God's sight, from God's sight to his having us "with" him, from his having us "with" him to his utterly unique manner of existence. (By "manner of existence" Augustine may have in mind specifically God's immutability, which he has just referenced.) Thus, God would not have created us if he did not exist uniquely as the one who is not made. Creation is not possible apart from divine priority.

Our current relationship to God is also grounded in divine priority. Later Augustine stipulates that God is both "interior to every single thing . . . and exterior to every single thing," just as he is both "older than all things . . . and newer than all things."[65] In this context Augustine quotes Acts 17:28, "For in him we live and move and are," affirming that the Creator's nearness to creatures surpasses their nearness to each other. Thus, once again, God's ontological priority

---

[62]*De Genesi ad litteram* 5.15.33 (CSEL 28:1, 158-59).
[63]*De Genesi ad litteram* 5.16.34 (CSEL 28:1, 159).
[64]*De Genesi ad litteram* 5.16.34 (CSEL 28:1, 159). Augustine proceeds to make clear that our preexistence as God's ideas is freely determined by God, and that it is the cause (not the effect) of God's work of creation.
[65]*De Genesi ad litteram* 8.26.48 (CSEL 28:1, 265).

does not slice him off from creation. Instead, God can be truly known and apprehended through his creatures. All things reflect him. As Vivian Boland puts it, "Each single thing is, in some way, a trace of God. . . . The being and therefore the truth of this world and its history cannot be understood without reference to the wise love of God which originates and sustains it."[66]

CREATURELY CONTINGENCY AND THE "CONVERSION TORQUE" OF CREATION

Divine priority entailed for Augustine what we will call *creaturely contingency*, the view that created reality, as a consequence of its participatory character, stands in a continual, dynamic orientation toward God. Simply stated: because *Creator* and *created* are the only two possible kinds of being, the nature and goal of creatures is radically dependent on the Creator. One is either Maker or made, and the latter's existence is completely oriented toward the former. As Jared Ortiz helpfully puts it, "For Augustine, creation has a 'conversion torque,' a dynamic orientation toward God, indeed, toward salvation."[67]

Creaturely contingency is the logical outgrowth of God's ontological priority, although the two state different strands of a paradox: Divine priority emphasizes the supreme distance between God and creation; creaturely contingency places them in a kind of intimate proximity. Divine priority entails that God is always beyond us; creaturely contingency entails that we are only ever in him. Divine priority means that God is infinitely alien; creaturely contingency means that he is our constant food and source.

Creaturely contingency, no less than divine priority, is bound up with creation *ex nihilo*: to pass into existence is to be contingent. Creaturely existence therefore has an inherently shared, derivative quality to it. Creatures *participate* in being, while God *is* being itself.

---

[66]As quoted in Levering, *Engaging the Doctrine of Creation*, 33.
[67]Ortiz, "*You Made Us for Yourself*," 230.

Thus, creaturely contingency is not an arbitrary sentence imposed by God onto creatures, but a necessary implication of God's identity as ontological source. In his *Confessions* Augustine refers to God as "Being itself" (*idipsum*) and claims that he does "not exist in a certain way, but he is, is" (*sed est, est*).[68] Creatures, by contrast, stand in that curious middle ground between God and nothing. As Joseph Torchia puts it, "Creatures occupy a mid-rank which situates them between the plenitude of Being found in God and absolute negation."[69] But in a sense, they are far "closer" to nothing than to God. "Creatures are inextricably bound up with non-being by virtue of their origins from nothing."[70] Thus, creatures are radically dependent on God for their continued existence. As Augustine puts it in his literal commentary on Genesis, "Anything, you see, that comes from him does so in such a way that it owes its being what it is to him."[71]

But this did not entail that creatures exist in a radically egalitarian position in relation to each other. Instead, Augustine's vision of created reality is hierarchical. On the ladder from nothing to divine plenitude, different creatures stand in a different rank, which is to say that some creatures exist more than others. As Augustine observes, "To some He communicated a more ample, to others a more limited existence, and thus arranged the natures of beings in ranks."[72] For Augustine, the human soul has a relatively high placement within his hierarchy, ranking alongside the angels as part of the "intelligent" creation, above mere physical creatures. As he once put it in a letter, "The soul is situated, of course, in a certain mid-rank, having beneath it the bodily creature but having above it the creator of itself and of its body."[73] On the whole, Augustine thinks of human nature as a "mean between the

[68]*Confessiones* 13.31 (CSEL 33, 383); cf. the discussion in Ortiz, *"You Made Us for Yourself,"* 6.
[69]Torchia, *Creatio Ex Nihilo*, 236.
[70]Torchia, *Creatio Ex Nihilo*, 165.
[71]*De Genesi ad litteram* 4.15.26 (CSEL 28:1, 112).
[72]*De civitate Dei* 12.2 (CSEL 40.1, 569).
[73]Letter 140, in *Letters (Epistulae) 100–155*, ed. Boniface Ramsey, trans. Roland Teske, The Works of St. Augustine: A Translation for the 21st Century (Hyde Park, NY: New City Press, 2003), 246.

angelic and bestial," capable of passing into the company of angels through obedience or descending into an animal-like existence through disobedience.[74] Thus, in the work of salvation, human nature is "united by the Holy Ghost to the holy angels in eternal peace."[75]

To understand creaturely contingency more fully, we must explore the point of contrast between Creator and creatures that Augustine emphasizes again and again in his writings on creation: divine immutability.[76] In his literal commentary Augustine distinguishes God's initial work of creation and his subsequent work of sustenance, arguing that "God is working until now in such a way that if his working were to be withheld from the things he has set up, they would simply collapse."[77] The reason for this inherent tendency toward non-being is that created things possess a mutable nature, not an immutable one. As Torchia notes, "If created being is drawn out of nothing by God, then it is inherently mutable, corruptible, and liable to pass out of existence. By virtue of its radical contingency, creation as a whole requires the continual support of God if it is not to degenerate into the utter non-being from which it was generated."[78] In his older study, Christopher O'Toole put it more simply: creation *ex nihilo* "implies on the part of the creatures a tendency to the nothingness from which they were drawn."[79]

Where in Augustine's writings do we see this? In his *De vera religione*, a treatise written soon after his conversion, Augustine identifies divine immutability as central not only to God's creative work, but also his providential work of sustaining the world: "If [God], after all, did not abide unchanging, no changeable nature would remain in

[74]*De civitate Dei* 12.21 (CSEL 40.1, 607).
[75]*De civitate Dei* 12.22 (CSEL 40.1, 607-8).
[76]E.g., *De civitate Dei* 11.10 (CSEL 40.1, 528-29).
[77]*De Genesi ad litteram* 5.20.40 (CSEL 28:1, 164).
[78]Torchia, *Creatio Ex Nihilo*, xii.
[79]Christopher J. O'Toole, *The Philosophy of Creation in the Writings of St. Augustine* (Washington, DC: Catholic University of America Press, 1944), 6-7.

existence at all."[80] Later, in the context of his argument that both the Old Testament and the New Testament come from the same source, Augustine develops this idea. After stipulating that the same doctor can prescribe different medicines to weaker and stronger patients, Augustine then intertwines this doctor/patient analogy with the immutable/mutable contrast between God and his creation:

> Divine providence, while being in itself absolutely unchanging, nonetheless comes to the aid of changeable creatures in various ways, and in accordance with the diversity of diseases commands or forbids different regimes—this in order to bring back from the malady which is the beginning of death, and from death itself, to their proper condition and state of being, and to strengthen them in it, the creatures that are failing, slipping, that is to say, into nothingness.[81]

Augustine here regards "changeable creatures" as in danger of "slipping . . . into nothingness," and therefore in need of the healing that immutable divine providence can provide.

Augustine then considers the question of why creatures are in this state in the first place, again turning on the immutability/mutability contrast:

> But you say to me: "Why are they failing (*deficiunt*)?"
> Because they are subject to change.
> "Why are they subject to change?"
> Because they do not have being in the supreme degree.
> "Why not?"
> Because they are inferior to the one by whom they were made.
> "Who is it that made them?"
> The one who *is* in the supreme degree.
> "Who is that?"
> God, the unchanging Trinity, since he both made them through his supreme Wisdom and preserves them through his supreme Kindness.[82]

[80]*De vera religione* 10.18 (CSEL 77, 15).
[81]*De vera religione* 17.34 (CSEL 77, 24).
[82]*De vera religione* 18.35 (CSEL 77, 24-25).

In this passage Augustine develops a chain of logic for creaturely contingency: creatures are "failing" (dying) because they are mutable; they are mutable because they do not have supreme existence; and they lack supreme existence because of the Creator/creation distinction, which for Augustine is bound up with a mutable/immutable ontological divide. Thus, creaturely contingency is entailed by the doctrine of creation: that which is changing/failing must be preserved by the kindness of the unchanging Trinity because it was brought into existence in the first place by his wisdom. Indeed, the very purpose of creation is that creatures would share in his being—as Augustine explains as he continues:

> "Why did he make them?"
>
> So that they might be. Just being, after all, in whatever degree, is good, because the supreme Good is being in the supreme degree.
>
> "What did he make them out of?"
>
> From nothing, since whatever is must have some kind of specific look, however minimal.

Creation *ex nihilo* here becomes the terminus of this thread of reasoning: existence of any kind is good, but that which lacks supreme existence must have been summoned into existence from nothing. As a whole, this passage highlights the significance of the divine immutability/creaturely mutability divide in the logic of creaturely contingency. Creatures are radically contingent upon God, most basically, because they are *mutable.*

Throughout Augustine's other writings as well, divine immutability is central to creaturely contingency because of creation *ex nihilo*. In his *De natura boni*, for example, Augustine identifies divine immutability as the chief point of distinction between the Creator and the creation because everything else is made from nothing: "He alone is immutable, while all the things that he has made are mutable because he has made them from nothing."[83] Similarly, in his *De libero arbitrio*, Augustine grounds creaturely contingency in his doctrine of creation *ex nihilo*: "As for temporal things, they have no existence before they exist; while they exist, they are passing away; once they have passed away, they will never exist again."[84] Augustine seems to envision that things which are made from nothing have a kind of inherently unstable and ephemeral quality of existence. They are continually "passing away." As he puts it just later, "beginning to exist is the same as proceeding toward non-existence."[85]

Now, why does Augustine consider divine immutability to play such a central role in this view of creaturely dependence on God? For Augustine, immutability is an integral part of God's perfection, since any change implies imperfection (otherwise there would be no need for change). This is why Augustine often associates mutability with becoming "defective" (*deficere*), and uses this is as a point of contrast between Creator and creation: "This I do know, that the nature of God can never, nowhere, nowise, be defective, and that natures made of nothing can."[86] Augustine also conceives of God's supreme existence as dependent on his immutability: "He truly is because he is immutable. Every change, after all, makes that which was not to be. He who is immutable, then, truly is."[87] Augustine's interest in immutability also comes in the context of his opposition to Manichaeism, predicated on his dual commitment to creation *ex nihilo* and the goodness of creatures. Thus, in his *Retractations*, Augustine summarized the purpose of his initial allegorical work on Genesis as showing "that God is supremely good and unchangeable, and yet the creator of all changeable natures, and that no nature or substance is evil precisely as a nature or substance."[88] Augustine is eager to counter the Manichaeans

---

[83]*De natura boni* 1 (CSEL 25.2, 855).
[84]*De libero arbitrio* 3.7 (CSEL 74, 107).
[85]*De libero arbitrio* 3.7 (CSEL 74, 107).
[86]*De civitate Dei* 12.8 (CSEL 40.1, 578).
[87]*De natura boni* 19 (CSEL 25.2, 863).
[88]*Retractationes* 1.9 (CSEL 36, 47).

by insisting that created things are good in themselves, but this concern with theodicy is bound up with his effort to contrast divine immutability and creaturely mutability.

Nonetheless, for all his use of the immutable/mutable contrast as a point of distinction between God and creatures, Augustine conceives of a kind of immutability that can be acquired by creatures as they share in God's life, tantamount to the highest state of human freedom: *non posse peccare*.[89] This is why it is important to Augustine to maintain that the good angels were confirmed in obedience when they did not join those angels who turned away from the Creator and fell: "To the most excellent creatures, that is, to rational spirits, God has granted that they cannot be corrupted if they are unwilling, that is, if they preserve obedience under the Lord, their God, and in that way cling to his incorruptible beauty."[90] We will return to this point below.

In Augustine's vision, then, human beings have as much to do with God at each moment of their existence as they do at the moment of their creation. We receive each breath as freely as we receive our very being; our need to turn to God continually is of the same quality as our need to have received existence from him in the first place. In his literal commentary on Genesis, Augustine correlates our once-for-all created status with our continual need of God's help in our moral formation: "Human beings, after all, are not the sort of things that, once made and left to themselves by the one who made them, could do anything well all by themselves. No, the sum total of their good is to turn to him by whom they were made, and by him always to be made just, godfearing, wise and blessed."[91] Human beings cannot become just or virtuous by God's help and then depart from him,

[89]As developed in *The City of God*, Augustine holds to four essential states in human nature: (a) able to sin, able not to sin (*posse peccare, posse non peccare*); (b) not able not to sin (*non posse non peccare*); (c) able not to sin (*posse non peccare*); and (d) unable to sin (*non posse peccare*). For Augustine, these states correspond to human nature before the fall, after the fall, after regeneration, and after glorification, respectively.

[90]*De natura boni* 7 (CSEL 25.2, 858).

[91]*De Genesi ad litteram* 8.12.25 (CSEL 28:1, 249).

having gotten what they needed—as a patient leaves a physician once cured, or a farmer leaves a field once he has finished cultivating it.[92] Rather, just as the sky is lit up by the continual presence of light, "in the same way human beings are enlightened by the presence of God with them, but immediately relapse into darkness with the absence of him from whom one distances oneself not by local movement but by the turning away of the will."[93] This third metaphor (light/darkness for divine presence/absence) emphasizes the perpetual, constant human need of God: as quickly as the darkness comes when there is no light, so quickly do human beings degenerate without God. It is not as though we slowly "fade out" when we lack God. There is simply no existence or goodness whatsoever apart from his sustaining support and help.

Augustine then emphasizes our continual dependence on God so strongly that there is even a sense in which we are *continually created* by him, right alongside our continual need for his help: "We ought always to go on being made by him, always being perfected by him, sticking to him and persevering in that way of life which is directed towards him."[94] Augustine emphasizes further in this context the link between our creation and our moral formation—we need God equally to be, and to be good. Our dependence on God is for both being and well-being. As he puts it later, commenting on 1 Corinthians 1:31: "'The one who glories should glory only in the Lord,' when he comes to know that it is not from himself but from God that he gets his very being; and what is more, that his well-being also comes only from the one from whom is derived his mere being."[95] Even when we are unaware of our continual need for God, we are no less conditioned by it. As he reflects in the *Confessions*, "Even when

[92]*De Genesi ad litteram* 8.12.25-26 (CSEL 28:1, 249-50).
[93]*De Genesi ad litteram* 8.12.26 (CSEL 28:1, 250).
[94]*De Genesi ad litteram* 8.12.27 (CSEL 28:1, 250).
[95]*De Genesi ad litteram* 11.8.10 (CSEL 28:1, 341).

all is well with me, what am I but a creature suckled on your milk and feeding on yourself, the food that never perishes?"[96] The ease with which Augustine moves back and forth between ontological and moral dimensions of our dependence on God testifies to the fact that his perception of creaturely contingency is not a merely speculative concern but a pastoral one. Hence it is not surprising to hear Augustine return to these same themes throughout his sermons, making appeals as direct as the following: "We owe it to God that we are what we are. From whom, if not from God, do we get it that we are not nothing?"[97]

Above all, Augustine was vitally interested in creaturely contingency because it concerned the happiness/beatitude of both humans and angels. Creaturely happiness is not often considered a theologically decisive topic, but it was central to Augustine's theology. It is not for nothing that Étienne Gilson opened his classic study of Augustine's philosophy with a discussion of human happiness (or beatitude) as consisting in the possession of God.[98] Augustine regarded our dependence on God for happiness as, once again, a consequence of creation *ex nihilo*. In *The City of God* he recognizes that many created things do not have the capacity to be "blessed" (*beatus*): "beasts, trees, stones, and things of this kind."[99] But he stipulates that the creature that *does* have this capacity (say, an angel) "cannot be blessed of itself, since it is created out of nothing."[100] Yet again, Augustine grounds this creaturely need on the distinction between God's immutability and creaturely mutability.[101]

---

[96]*Confessiones* 4.1 (CSEL 33, 64).

[97]Sermon 43, in *Sermons 20–50 on the Old Testament,* ed. John E. Rotelle, trans. Edmund Hill, The Works of Saint Augustine: A Translation for the 21st Century (Hyde Park, NY: New City Press, 1992), 271.

[98]Étienne Gilson, *The Christian Philosophy of Saint Augustine*, trans. L. E. M. Lynch (New York: Random House, 1960), 3-10.

[99]*De civitate Dei* 12.1 (CSEL 40.1, 567).

[100]*De civitate Dei* 12.1 (CSEL 40.1, 567).

[101]*De civitate Dei* 12.1 (CSEL 40.1, 567).

TRINITARIAN AGENCY IN CREATION

Augustine develops his conception of creaturely contingency in relation to God's trinitarian agency in creation. Augustine speaks of the "unformed" state of created things, by which he means their propensity to slip into nonexistence unless they adhere to the divine Word by participating in the union between the Father and the Son. He conceives of this act as creaturely *formation*, logically distinct from *ex nihilo* creation, and as a kind of "conversion" (*conversio*). Now, it must be remembered that the acts of creation and formation cannot be divided from one another. Thus, Étienne Gilson claimed that for Augustine, "The divine act of creation combines in its indivisible unity the production of two different effects. . . . To create, then, means that God, by one and the same act, produces the unformed and calls it back to Himself so as to form it."[102] Nonetheless, creation and formation can be distinguished, and they pattern trinitarian dynamics within the Godhead.

In his older study of Augustine's doctrine of creation, Christopher O'Toole summarized Augustine's teaching on this point as follows:

> Matter is dissimilar to that which supremely is (*summe ac primitus est*). Because of that it possesses a natural tendency to non-existence (*informitate quadam tendit in nihilum*). In this it does not imitate the form which is the Word through which God speaks from all eternity. But it does imitate the Word (*Form*) always and unchangeably adhering to the Father when by a turning (*conversio*) to that which truly and always is, i.e., the Creator of its substance, it receives a form and becomes a perfect creature. The formation, therefore, both of spiritual and corporeal matter involves a *conversio*.[103]

This is a helpful summary, although it does not mention the role of the Holy Spirit in the formation of creatures. More recently Jared

[102]Gilson, *Christian Philosophy of Saint Augustine*, 205.
[103]O'Toole, *Philosophy of Creation*, 22.

Ortiz, drawing on the work of Marie-Anne Vannier, suggests that Augustine uses the terms *creatio*, *conversio/revocatio*, and *formatio* to describe the activities of the Father, Son, and Spirit, respectively.[104] Thus, the Father summons each thing into being from nothing (*creatio*) into an unformed state; he then calls this unformed thing back to himself through his Word, or Son (*conversio/revocatio*); the Holy Spirit then forms each thing according to the Father's speech (*formatio*).[105] These different aspects of trinitarian agency in creation are, again, ultimately inseparable, yet Augustine does appropriate them among the persons of the Godhead. Thus, in Augustine's exegesis of Genesis 1, the Son's work of conversion corresponds to the statements "let there be," and the Spirit's work of formation corresponds to the statements "and there was" (more on this in chapter 3).

Now, this is not the language Augustine always employs or the only language he uses. But it is striking how consistently he develops his conception of creation in terms of this broad trinitarian schema, with the implication that creation possesses a dynamic orientation toward God. In the act of creation, the triune God not only brings into being that which was not there before, but converts or recalls it to himself, forming it into the being he intends for it (which occurs through participation in himself). This can be seen, for instance, in a significant passage from the early chapters of Augustine's literal commentary on Genesis. He is inquiring about the "unformed basic material" of Genesis 1:2, and why the text does not record the phrase "and God said, 'let it be made'" of this material, as it does of the subsequent created things in 1:3–2:3.[106] As a possible answer, Augustine envisions God the Son as the divine speech/wisdom/light by which God creates

[104]Ortiz, "*You Made Us for Yourself*," 12.

[105]As Fox, *Augustine*, 7, notes, the term *conversio* "could refer to God's own 'turning' to His universe or to individuals in it. This sense, a turning 'towards' us by God, is the sense of the word *conversio* when it first occurs in the *Confessions*. In their concluding meditation on Genesis, turning, or *conversio*, is the first turning of formless matter, the basic constituent of the universe. It 'converts' away from dark chaos towards God's light."

[106]*De Genesi ad litteram* 1.4.9 (CSEL 28:1, 7-8).

throughout Genesis 1—thus, the phrase "let it be made" refers neither to "the sound of a voice nor thoughts running through the time which sounds take," but rather to "the light, co-eternal with [God], of the Wisdom he has begotten."[107] After all, Augustine reasons, the unformed basic material of Genesis 1:2 is imperfect, and "imperfection or incompleteness does not imitate the form of the Word, being unlike that which supremely and originally is, and tending by its very want of form toward nothing."[108]

We are by now well familiar with the ontological contrast being drawn here, but the reference to imitation of the form of the Word introduces a new layer. Augustine continues by explaining that the final perfection for which each creature is designed, that second goal or terminus that we have previously identified, occurs through an imitation of the Son's relationship to the Father:

> It is when it turns, everything in the way suited to its kind, to that which truly and always is, to the creator that is to say of its own being, that it really imitates the form of the Word which always and unchangingly adheres to the Father, and receives its own form and becomes a perfect, complete creature. . . . By so turning back and being formed creation imitates, every element in its own way, God the Word, that is the Son of God who always adheres to the Father.[109]

Through his writings on creation, Augustine will often speak of creation *ex nihilo* as a kind of foil or contrast to the eternal generation of the Son: thus, God begets the Son; God creates the world; and although these are two distinct kinds of relation, the latter bears some faint analogy of the former.[110] Often the contrast between *begottenness* and *ex nihilo* serves to safeguard the superiority of Creator over creation, since what God begets from himself has priority over what he

---

[107]*De Genesi ad litteram* 1.4.9 (CSEL 28:1, 7).

[108]*De Genesi ad litteram* 1.4.9 (CSEL 28:1, 7).

[109]*De Genesi ad litteram* 1.4.9 (CSEL 28:1, 7-8).

[110]E.g., *De Genesi contra Manichaeos* 1.2.4 (CSEL 91, 70); *De Genesi ad litteram liber unus imperfectus* 5.20 (CSEL 28:1, 472); *De civitate Dei* 11.10 (CSEL 40.1, 528-29).

makes from nothing. For instance, early on in his allegorical commentary he claims, "Nor did [God] beget [good things] from himself, to be what he is himself, but he made them out of nothing, so that they would not be equal either to him by whom they were made, or to his Son through whom they were made."[111]

But in the passage just quoted above (*De Genesi ad litteram* 1.4.9), Augustine goes one step further. Not only does the creation of creatures from nothing bear some kind of analogy to the Son's begottenness from the Father, but the formation and perfection of creatures bears some kind of analogy to the Son's "adhering" (*cohaerens*) to the Father. Thus, not only do created things have an inherent disposition toward decay into nothingness (creaturely contingency); and not only must they receive their final, perfected form by clinging to the One who possesses true and supreme existence (divine priority); but the *way* they cling to him is by imitating how the Word clings to the Father. Augustine's vision of *creaturely* orientation to God is thus patterned after trinitarian dynamics: creaturely reality becomes "perfect" by imitating the Son. Moreover, the Son of God is active in this process, not merely an example to be imitated. Thus, a few sentences later Augustine references the Son's role of "conferring perfection on creation by calling it back to himself, so that it may be given form by adhering to the creator, and by imitating in its own measure the form which adheres eternally and unchangeably to the Father, and which instantly gets from him to be the same thing as he is."[112] It is worth noting here that Augustine does not perceive this process of adhering to the Father to obliterate creaturely distinctiveness—each created thing adheres to the Father "in its own measure." This also recalls his assertion just earlier that each creature turns to God in a way "suited to its kind."[113]

---

[111]*De Genesi contra Manichaeos* 1.2.4 (CSEL 91, 70).
[112]*De Genesi ad litteram* 1.4.9 (CSEL 28:1, 8).
[113]*De Genesi ad litteram* 1.4.9 (CSEL 28:1, 8).

The active role of the divine Son in calling creatures to "formation" is also emphasized in the *Confessions*. In his treatment of Genesis 1 in this book, Augustine correlates the water of Genesis 1:2 with the spiritual creation in its incipient ("unformed") state, declaring that it "was like the depths of the ocean and it would have remained in that state, estranged from your likeness, unless that same Word had turned it towards its Creator and made it light by casting his own brightness upon it, not in equal degree with yourself, but allowing it to take form in your likeness."[114] Augustine continues: "The good of the spirit is to cling to you forever, so that it may not, by turning away from you, lose the light which it gained by turning towards you and relapse into that existence which resembles the dark depths of the sea." Once again, creaturely "clinging" to God is both an imitation of the Son and an activity to which the Son himself summons creatures, shining his light on them.

Augustine wants, of course, to preserve an absolute distinction between the way in which "formless" (*informis*) created things adhere to the Father and the way the Son adheres to the Father. To this end he draws on the doctrine of divine simplicity. As he stipulates, the Son "does not have an unformed life, seeing that for him not only is it the same thing to be as to live, but to live is for him the same as to live wisely and blessedly."[115] Created things, by contrast, may possess existence without possessing wise or happy existence, and thus must be "formed" by "turning to the unchangeable light of Wisdom, the Word of God."[116] Because the divine Son inherently possesses these qualities, he never ceases to speak to and summon each creature to its formation in the One from whom it derived its being, since there is no other way for a creature to achieve this end. In other passages, Augustine emphasizes the Holy Spirit's role in the formation of created

[114]*Confessiones* 13.2 (CSEL 33, 346).
[115]*De Genesi ad litteram* 1.5.10 (CSEL 28:1, 8).
[116]*De Genesi ad litteram* 1.5.10 (CSEL 28:1, 9).

things. For instance, at one point in his literal commentary Augustine references a Syrian rendering of the Spirit's activity in Genesis 1:2 as "brooding" (Latin *fovebat*) over the water "in the way birds brood over their eggs, where that warmth of the mother's body in some way also supports the forming of the chicks through a kind of influence of her own love."[117] Augustine does not hesitate to affirm, in this context, that in creation God works "by the eternal and unchanging, stable formulae of his Word, co-eternal with himself, and by a kind of brooding, if I may so put it, of his equally co-eternal Holy Spirit."[118] In the *Confessions* also, Augustine emphasizes the Spirit's role in these dynamics of creation.[119]

What emerges from all this is how Augustine situates the participatory, rather than autonomous, character of created existence in relation to the trinitarian dynamics within God. Each created thing is not only called into being from nothing by the Father, but stands in need of a further formation by the Son (and Spirit) in order to find its true nature in union with God in his immutable "rest." But this raises the question: If creatures are already radically contingent on a divine *conversio* simply as a result of their creaturely status, what new situation does sin introduce?

SIN AS PRIVATION

The notion of evil as privation, which Augustine developed for his own theological ends, had been clearly articulated by earlier philosophers such as Plotinus and Epictetus.[120] But Augustine made this notion a pillar of his ontology. As G. R. Evans notes, Augustine consistently used vocabulary to describe evil that connotes a falling away

[117]*De Genesi ad litteram* 1.18.36 (CSEL 28:1, 26-27).

[118]*De Genesi ad litteram* 1.18.36 (CSEL 28:1, 26). In *On Genesis*, ed. John E. Rotelle, trans. Edmund Hill, The Works of Saint Augustine: A Translation for the 21st Century (Hyde Park, NY: New City Press, 2002), Hill notes the reference here is likely to St. Ephrem (185).

[119]*Confessiones* 13.5-7 (CSEL 33, 348-50).

[120]See the discussion in Evans, *Augustine on Evil*, 2.

or turning away from good.[121] In his *Enchiridion*, which he intends as a summary of Christian belief, Augustine asks, "What else is that which is called evil but a removal of good?"[122] Augustine further maintains that if good did not exist, there could be no such thing as evil, and that nothing can ever be *wholly* evil, since if it were it would cease to exist.[123]

Stemming from his dispute with the Manichaeans, it was important for Augustine to make clear that there is no such thing as an evil nature. In *De natura boni*, he affirms that evil does not consist in desiring a bad nature but abandoning a better one.[124] He therefore claims, "The act itself is evil, not the nature of which one makes bad use in sinning."[125] Similarly, in *The City of God*, Augustine insists that even the fallen angels have a nature derived from God, and not some other source, since there is no essence contrary to the divine: "To that nature which supremely is, and which created all else that exists, no nature is contrary save that which does not exist."[126] As a consequence of this, Augustine argues that evil is not injurious to God, but only to itself.[127]

Because evil is essentially privative, there is even a sense in which it derives its existence from mimicking good. For example, Augustine regards pride as, despite itself, essentially imitative of God. As he puts it in the *Confessions*, "Even those who go from You and stand up against You are still perversely imitating You. But by the mere fact of their imitation, they declare that You are the creator of all that is, and that there is nowhere for them to go where You are not."[128] Thus, the act of turning away from the Creator is bad—but the very ability to do so is good. To be a *self* entails the possibility of

[121]Evans, *Augustine on Evil*, 95, draws attention specifically to Augustine's use of the terms *perversus, perversitas, aversio, defectio, lapsus, deformitas, deviare,* and *infirmare*.
[122]*Enchiridion* 3.11, in *On Christian Belief*, 278.
[123]*Enchiridion* 4.12-13, in *On Christian Belief*, 280.
[124]*De natura boni* 34 (CSEL 25.2, 872).
[125]*De natura boni* 36 (CSEL 25.2, 873).
[126]*De civitate Dei* 12.2 (CSEL 40.1, 569).
[127]*De civitate Dei* 12.3 (CSEL 40.1, 570-71).
[128]*Confessiones* 2.6 (CSEL 33, 40); translation from Sheed, *Confessions*, 32.

misusing one's constituent faculties (mind, will, etc.). But this very misuse testifies to the goodness of these faculties and consequently reveals God's inescapability.

Augustine regards sin as inherently involving a movement from a greater to a lesser degree of being since it consists of a movement away from God, the source of all being. Thus in *De vera religione* he stipulates that life is found in God because he is the Supreme Being and the fountain of life, while sin involves a movement away from God toward nothing. Those who experience total death cease to exist altogether, while those who progress toward death "participate less in being."[129] This draws from Augustine's Platonic ontology, in which a creature's degree of being is directly correlated to its closeness to God: "All things have their origin in you, whatever the degree of their being, although the less they are like you, the farther they are from you, and here I am not speaking in terms of space."[130] Thus, Augustine holds that at the fall, Adam and Eve did not cease to exist, but rather by turning inward away from God they became more "contracted" than they had been, since their relation to God was severed. As he puts it in *The City of God*, "Man did not so fall away as to become absolutely nothing; but being turned toward himself, his being became more contracted than it was when he cleaved to him who supremely is."[131]

The presupposition of this kind of language is that sin has a kind of logical trajectory toward non-being—a kind of ontological downgrading—since it involves a movement away from God, the source of all life and being. Augustine does not set this ontological consequence of sin at odds with its moral character. Rather, sin is destructive precisely because it is evil. As Matthew Levering puts it, "Sin lessens participation in God, and so is hateful and destructive."[132] Augustine also maintains that God actively punishes sin—but his ontological

[129] *De vera religione* 11.22 (CSEL 77, 17).
[130] *Confessiones* 12.7 (CSEL 33, 314).
[131] *De civitate Dei* 14.13 (CSEL 40.2, 32).
[132] Levering, *Theology of Augustine*, 128.

framework entails that God's punishment of sin is not the enforcing of an arbitrary rule, but the inevitable consequence to what sin is. Similarly, the final redemption of creatures is not an arbitrary bestowing of life but a participation in the God who *is* life.[133]

Thus, importantly, Augustine holds that evil is not merely a privation of being but a destruction of being. For example, in the spring of 391, soon after his ordination as a bishop in Hippo, Augustine preached a sermon "handing over" the (Apostles') creed to several catechumens, who would then "give back" the creed the following week.[134] Augustine spends considerable time at the start of the sermon emphasizing the privative nature of evil—he does not want his catechumens harboring any Manichaean suspicions that evil is a "something" lurking out there in the universe, a positive entity set at odds with good. Then, after reminding his listeners that they must defend the creed from its opponents, he stipulates, "So remember that there is absolutely no nature which he did not create. And that's why he punishes sin, which he didn't make, because it fouls nature, which he did create."[135] Here Augustine goes beyond merely claiming that God did not create evil and that evil is nothing in itself; God punishes evil because it "fouls" nature. Thus, for Augustine, evil is not only something that is not made; it is that which unmakes. It is not only a gap but an acid. Augustine then develops his comments further along the lines of a theodicy, arguing that the wicked have no right to believe that their defying God's will constrains his omnipotence because God uses their evil to accomplish a greater good.[136] Still, Augustine is

[133]For a striking instance of Augustine's participatory ontology, see his discussion of chastity and wisdom in relation to the *imago Dei* in *De Genesi ad litteram liber unus imperfectus* 16.57 (CSEL 28:1, 498).

[134]In *Sermons 184–229Z on the Liturgical Seasons*, ed. John Rotelle, trans. Edmund Hill, The Works of Augustine: A Translation for the 21st Century (Hyde Park, NY: New City Press, 1993), John Rotelle notes that there is some dispute about the dating of the sermon and suggests that Augustine revised it and preached it again later in life (157-58).

[135]Sermon 214.2, in *Sermons 184–229Z*, 151.

[136]Sermon 214.3, in *Sermons 184–229Z*, 152.

careful to distinguish between what God *allows* and what he *creates*, and again casts sin as anti-nature: "What he created was their natures, not their sins which are against nature."[137]

All this Augustine draws from his exposition of the phrase, "I believe in God, the Father Almighty, creator of heaven and earth," and in the context of a sermon preparing catechumens for baptism. In fact, the energy that Augustine devotes to developing the doctrine of creation here far surpasses his exposition of the rest of the creed. This highlights again the extent to which Augustine regarded the doctrine of creation as a pastoral and not merely speculative concern, as well as the importance of the doctrine of creation for grounding a proper understanding of good and evil.

To sum up, Augustine regards sin as leading to non-being because it severs the creature's dependence on God, which is necessary for its being and well-being. Thus, Augustine's conception of sin draws on his doctrine of creaturely contingency, grounded in his conception of God as the source of all being. How, then, does God's redemptive activity remedy this situation?

REDEMPTION AS DEIFICATION

If evil involves a movement down the hierarchical chain of being into nothingness, redemption for Augustine consists of ascending up into God as the source of all being—what is sometimes called deification, or divinization. The notion of deification has often been regarded as an Eastern concept, in contrast to the prominence of legal categories in the Western tradition, but in fact deification is widely represented in the Western tradition as well.[138] It is clearly a part of Augustine's thought and coheres naturally with his emphases on creaturely contingency, evil

---

[137]Sermon 214.3, in *Sermons 184–229Z*, 152.

[138]E.g., see Carl Mosser, "An Exotic Flower? Calvin and the Patristic Doctrine of Deification," in *Reformation Faith: Exegesis and Theology in the Protestant Reformations*, ed. Michael Parsons, (Milton Keynes: Paternoster, 2014), 38-56; Mosser, "Deification: A Truly Ecumenical Concept," *Perspectives: A Journal of Reformed Thought* 30 (4): 8-14.

as privation, and God as the source of all being. Yet as Bonner notes, despite its debt to a Platonic ontology, Augustine's notion of deification is distinguished by its emphasis on the incarnation.[139] To convey the notion of deification, Augustine used a wide variety of images and terms, including identification, assimilation, recapitulation, adoption, divine exchange, and participation.[140] This language is drawn from an array of biblical texts, though Augustine steers away from 2 Peter 1:4 because of its misuse to promote the Pelagian doctrine of human perfection.[141] As hinted at above, he also casts deification as a participation in God's immutability—for instance, in *The City of God* he speaks of the beatific vision as a "participation in (God's) unchangeable immortality, which we burn to attain."[142]

There are many passages in Augustine to which we might turn to develop his teaching on this subject.[143] Here let us just consider one recently discovered source. In 1990 the French medievalist François Dolbeau stumbled upon some twenty-six previously lost sermons of Augustine in the stacks at the Stadtbibliothek in Mainz, Germany, collated sometime in the fifteenth century.[144] One of them, a sermon on Psalm 81, opens by identifying deification as the Christian hope:

> To what hope the Lord has called us, what we now carry about with us, what we endure, what we look forward to, is well known, I don't doubt, to your graces. We carry mortality about with us, we endure infirmity, we look forward to divinity. For God wishes not only to vivify, but also to deify us.[145]

---

[139]Gerald Bonner, *St. Augustine of Hippo: Life and Controversies*, 3rd ed. (Norwich: Canterbury, 2002), 4-5.

[140]See the discussion in David Vincent Meconi, "Augustine's Doctrine of Deification," in *The Cambridge Companion to Augustine*, 215-21; see also Meconi, *The One Christ: St. Augustine's Theology of Deification* (Washington, DC: Catholic University of America Press, 2013).

[141]Meconi, "Augustine's Doctrine of Deification," 225.

[142]*De civitate Dei* 12.20 (CSEL 40.1, 602).

[143]For a fuller treatment, and a list of quotes from Augustine expressing his understanding of deification, see Gerald Bonner, "Deification, Divinization" in *Augustine Through the Ages*, 265-66.

[144]See the discussion in Meconi, "Augustine's Doctrine of Deification," 216.

[145]Sermon 23B.1, in *Sermons Discovered Since 1990*, ed. John E. Rotelle, trans. Edmund Hill, The Works of Saint Augustine: A Translation for the 21st Century (Hyde Park, NY: New City Press,

As the sermon continues, Augustine develops the notion of deification in terms of the divine exchange involved in the incarnation of the Word:

> We mustn't find it incredible, brothers and sisters, that human beings become gods, that is, that those who were human beings become gods. More incredible still is what has already been bestowed upon us, that one who was God should become a human being. And indeed we believe that that has already happened, while we wait for the other thing in the future. The Son of God became a son of man, in order to make sons of men into sons of God. . . . The maker of man was made man, so that we might be made a receiver of God.[146]

Augustine's language here bears some obvious similarities to famous statements about deification by Irenaeus and Athanasius, and like them he displays a concern to maintain a robust distinction between Creator and creature, and to steer clear of polytheism. He cautions that we are not of divine substance by nature any more than the Son of God was mortal before being united to humanity. The true God is God by nature; other gods are so by the grace of adoption. Indeed, God's very ability to deify others distinguishes his own deity: "There is a great difference between God who exists, God who is always God, true God, not only God but also deifying God; that is, if I may so put it, god-making God, God not made making gods, and gods who are made, but not by a craftsman."[147]

Thus, it is important to stress that Augustine did not consider deification to consist of any diminution of our creatureliness, finitude, or humanity. Deification is not the loss of our nature, but rather its fulfillment. Meconi defines it as "the perfection of the human as he or she comes to live in total and perfect union with God."[148] We would

---

1997), 37.

[146]Sermon 23B.1, in *Sermons Discovered Since 1990*, 37.

[147]Sermon 23B.2, in *Sermons Discovered Since 1990*, 38.

[148]Meconi, "Augustine's Doctrine of Deification," 225.

add that Augustine understood deification, more specifically, to be the fulfillment of the *imago Dei*. As he puts it in the *Soliloquies*, "Your image in a mirror wants, in a way, to be you and is false because it is not. . . . All pictures and replicas of that kind and all artists' works of that type strive to be that in whose likeness they are made."[149] Thus, deification for Augustine stresses continuity between creation and redemption: since we were made in God's image, we are incomplete and lacking until we are drawn into God himself.

For Augustine, it is fundamentally the incarnation that establishes the possibility for deification—in assuming human flesh, the Son of God becomes the mediator of divine-human union in which the original creational purpose of humanity can be accomplished.[150] It is the divine humility entailed in the incarnation, paralleled with his underived power to deify, that distinguishes the true God from the false "gods" of the pagan world.[151] Thus, in his literal commentary on Genesis, Augustine speaks of the Son of God's illuminating work as spreading into creation "a kind of infection (*adfectio*) of shining, glowing intelligence."[152] The incarnation has this soteriological significance in Augustine's thought because he identifies the Son of God as the Light and Wisdom of God through which the world was originally created.

As we have said, Augustine's conception of deification thus entails a strong link between creation and redemption. What is accomplished by redemption is not merely the undoing of sin, but also the achievement of our original creational purpose of imaging God. Therefore, for Augustine, redemption is not an intrusion into our creaturely status but its fulfillment. Furthermore, the radical need for God involved in redemption draws on, though it surpasses, the radical need for God *already* entailed by creation *ex nihilo*. Thus, it is

[149]As cited in Meconi, "Augustine's Doctrine of Deification," 213. These statements are put as questions in the dialogue of which they are a part, but in context are answered affirmatively.
[150]Cf. Ortiz, "*You Made Us for Yourself*," 110.
[151]Sermon 23B.3-4, in *Sermons Discovered Since 1990*, 38-39.
[152]*De Genesi ad litteram* 1.17.32 (CSEL 18:1, 24).

not as though redemption requires a posture of dependence on God whereas unfallen createdness had entailed an autonomous, independent status. Rather, redemption intensifies the dynamic orientation to God that is already implicit in createdness.

This observation has the inconvenience of standing in direct opposition to Colin Gunton's reading of Augustine. Gunton argues (among other points) that Augustine makes redemption out to be a kind of rescue from createdness. This, of course, is part of his broader critique in which many of the various ills of modern Western society are attributable to Augustine's chain of influence.[153] Gunton's interpretation of Augustine may be worth briefly engaging because in some respects it is representative of several broader trends in Augustine scholarship (and to some extent, patristic scholarship more broadly) that have generally fallen out of scholarly favor yet continue to crop up in many popular-level and survey texts. The first is a sharp divide between the Eastern (Greek) and Western (Latin) traditions in the early church on the Trinity, according to which the Eastern church emphasized God's threeness while the West emphasized his oneness. Augustine naturally plays a critical role in this historical sketch, since it is claimed that the Western tradition's conception of the Trinity was plagued by his overly philosophical construal of divine unity.[154] As Lewis Ayres has demonstrated, however, this East/West contrast is far less significant than commonly supposed, and Augustine's trinitarianism bears many essential points of continuity with that of Eastern counterparts.[155] For instance, Augustine emphasized the irreducibility

[153]E.g., Colin E. Gunton, *The Triune Creator: A Historical and Systematic Study*, New Studies in Constructive Theology (Grand Rapids: Eerdmans, 1998), 73-86. Throughout various writings Gunton critiques a number of other areas of Augustine's doctrine of creation, including his Platonic inheritance and the concern that he makes creation arbitrary.

[154]Such a view is too commonplace to document fully, but to identify one early text that influenced the scholarship in this respect in significant ways, see Théodore de Régnon, *Études de théologie positive sur la Sainte Trinité* (Paris: Retaux, 1898). For analysis of Régnon's text, see Michel René Barnes, "De Régnon Reconsidered," *Augustinian Studies* 26 (1995): 51-71.

[155]Lewis Ayres, *Augustine and the Trinity* (Cambridge: Cambridge University Press, 2010); see also Ayres, *Nicaea and Its Legacy: An Approach to Fourth-Century Trinitarian Theology* (Oxford: Oxford University Press, 2006); and Ayres, "The Question of Orthodoxy," *Journal of Early Christian*

of the three persons and construed divine unity in terms of the Father's eternal act producing a communion of love among the three persons.[156]

This strong Greek/Latin division is naturally bolstered by the popular notion that Augustine was cut off from Greek theology by his ignorance of the language. But this claim also is generally overblown or put to an overly ambitious use. While it is true that Augustine was always somewhat naturally averse to the language, he did gradually improve his understanding of Greek throughout his lifetime. During the Pelagian controversy, Augustine provided his own translation when quoting a Greek father; in 428, he translated an entire Greek treatise (St. Epiphanius's *Panarion*); already by about 415–416, Augustine had, according to Gerald Bonner, "a reasonable working knowledge of biblical Greek."[157] More basically, beyond considerations of linguistic skill, it is evident that Augustine was not ignorant of the Eastern tradition—his *On the Trinity*, for instance, demonstrates familiarity not only with the Latin fathers Tertullian and Hilary but also with Basil and Didymus the Blind (possibly through Ambrose), and possibly with the other Cappadocian fathers.[158] It is problematic to confuse a lack of skill at the Greek language with an ignorance of Greek theology.

A third plague upon Augustine's reputation in modern perception, corollary to the East/West-divide and ignorance-of-Greek theories, is the so-called Hellenization thesis, in which Christian theology is held to be corrupted by the influence of Greek philosophy, particularly Neoplatonism. Augustine generally features prominently in this historical interpretation.[159] In earlier generations, Adolf von Harnack

Studies 14.4 (2006): 395-96, where he describes the general scholarly movement away from the Hellenization thesis over the last fifty years.

[156]Ayres, *Augustine and the Trinity*, 319. Ayres mentions as an example Augustine's taking Nicaea's "God from God, Light from Light" language in terms of God's intra-trinitarian relations (3).

[157]Bonner, *St. Augustine of Hippo*, 394-95.

[158]Cf. Yves Congar, *I Believe in the Holy Spirit*, vol. 3, trans. David Smith (1979–1980; repr., New York: Crossroad, 2000), 80-81.

[159]E.g., Olivier Du Roy, *L'Intelligence de la foi en la Trinité selon saint Augustin: Genèse de sa théologie trinitaire jusqu'en 391* (Paris: Études augustiniènnes, 1966). For an appraisal of Du Roy's position, see Ayres, *Augustine and the Trinity*, 20-41.

popularized the Hellenization thesis, but many scholars today are urging that it be put to bed, and that the (complicated) relation of early Christianity to Greek thought is better described as the Christianization of Hellenism rather than the Hellenization of Christianity.[160]

Perhaps because Gunton's historical portrait of Augustine relies on both the Hellenization thesis and the East/West dichotomy, it tends to pit certain elements of Augustine's thought at odds with one another unnecessarily. For instance, far from conceiving redemption as the escape from creation, Augustine, as we have seen, regards deification as consisting of the fulfillment of our human nature, and specifically of our creation in God's image. Thus, as Bradley Green points out, in Augustine's view our created nature is precisely oriented to the end of the beatific vision.[161] Moreover, Augustine conceives of redemption as bringing to completion not only human nature, but the entire world. Thus, in his sermons he frequently draws on the notion of Sabbath rest as a type of heavenly contemplation of God,[162] and he specifies the end goal of creation as a return to its starting point of rest.[163] This way of thinking stems from his vision of all creation possessing a built-in goal of sharing in divine immutability, as we have observed, and so attaining the "goal of its momentum,"[164] when all creation will share in the divine rest of "the Sabbath of eternal life."[165]

---

[160]For further discussion, see Paul L. Gavrilyuk, *The Suffering of the Impassible God: The Dialectics of Patristic Thought*, Oxford Early Christian Studies (Oxford: Oxford University Press, 2006); Ayres, *Nicaea and Its Legacy*; Jaroslav Pelikan, *Christianity and Classical Culture: The Metamorphosis of Natural Theology in the Christian Encounter with Hellenism*, Gifford Lectures Series (New Haven, CT: Yale University Press, 1995).

[161]Bradley G. Green, *Colin Gunton and the Failure of Augustine: The Theology of Colin Gunton in Light of Augustine*, Distinguished Dissertations in Christian Theology 4 (Eugene, OR: Pickwick, 2011), 179-81.

[162]E.g., Sermon 251.5, in *Sermons 230–272B on Liturgical Seasons*, ed. John E. Rotelle, trans. Edmund Hill, The Works of Saint Augustine: A Translation for the 21st Century (Hyde Park, NY: New City Press, 1993), 129.

[163]Sermon 259, in *Sermons 230–272B on Liturgical Seasons*, 177-82. By the time he writes *The City of God*, Augustine will have adjusted his understanding of this point and interpret the millennium as the entire time intervening between the first and second coming of Christ (cf. *De civitate Dei* 20.7.1).

[164]*De Genesi ad litteram* 4.18.34 (CSEL 28:1, 117).

[165]*Confessiones* 13.36 (CSEL 33, 387).

WHAT CAN CONTEMPORARY CREATION THEOLOGY LEARN FROM AUGUSTINE'S ONTOLOGICAL VISION?

Much more could be said about the ontological substructure of Augustine's doctrine of creation and its resulting implications for creaturely beatitude, fallenness, and redemption. But even this brief overview can perhaps draw attention to some areas where evangelicals might stretch their own thinking about creation by engaging Augustine. After all, living abroad for a mere four months can already dramatically influence your perception of your own country. You need not become an expert before you can start learning.

Of course, few of us would adhere to Augustine's ontology in detail, and some of us may have sharp disagreements with him on certain points. Nonetheless, there are several ways that it might be instructive, particularly to the extent that creation played a generally broader role in his theology as a whole than it often does in the contemporary discussion.

So to return to our earlier metaphor: suppose representatives of Answers in Genesis, Reasons to Believe, and BioLogos are around the table having a conversation. Now Augustine enters. How might his presence alter and expand the conversation about creation? Here are three possibilities: Augustine may *refocus* us on the central issues, *remind* us of neglected topics, and *redirect* us toward unconsidered possibilities.

First, Augustine may influence evangelical treatment of creation by refocusing us on central issues within the doctrine of creation. That is to say, including Augustine's voice in the discussion may draw our attention back toward those matters in the doctrine of creation that are arguably most important to it, most agreed upon throughout the church (past and present), and most relevant to our witness and worship. Evangelical discussion of creation is highly polemical—we can hardly hear the term without thinking of the "creation wars." Many Christians are not even aware how much more has been at stake, historically, in the doctrine of creation beyond the areas of current dispute. Augustine can make us more aware of our common ground in vital areas such as creation *ex nihilo*, the *imago Dei*, and the goodness of creation.

Appreciating the ontological underpinnings of Augustine's doctrine of creation can help us appreciate, for instance, the uniqueness of a Christian way of looking at our world as distinct from its alternatives. As Ian McFarland notes, "Christian teaching about creation is more idiosyncratic than is generally realized."[166] Early on, even some prominent Christians (such as Justin Martyr) affirmed an account of creation that had more similarities to Plato's demiurge reshaping formless matter than to creation *ex nihilo*.[167] But eventually *ex nihilo* held sway. One of the reasons Christians fought for this truth is that it marked off a Christian way of construing the nature of the God-world relationship that preserved divine transcendence, in opposition to various gnostic and heterodox alternatives. Thus, for the early Christians, creation was less concerned with how long God took to create, and more concerned with whether you had the right kind of God doing the creating.

We often take this legacy for granted today. In the midst of Christian debates about when the world began, we often overlook our decisive affinity in believing *that* it had a beginning. It is unhelpful to emphasize the peripheral, disputed matters while simultaneously downplaying the agreed-upon, central matters. Sympathizing with Augustine's struggle against the Manichaeans, truly understanding what he was up against—this is one way to set our current disputes in context. To be sure, the differences between theistic evolution, old-earth creationism, and young-earth creationism are important and should not be minimized. But with respect to what has historically been the *most*

[166]Ian A. McFarland, *From Nothing: A Theology of Creation* (Louisville, KY: Westminster John Knox, 2014), xi.

[167]Justin Martyr, *First Apology* 59, in Ante-Nicene Fathers: The Apostolic Fathers with Justin Martyr and Irenaeus 10 vols., ed. Alexander Roberts and James Donaldson (1885; repr., Peabody, MA: Hendrickson, 2012), 1:182-83.

important issue—the nature of the God-world relationship—the world could be thirteen billion years old, or thirteen thousand, and it would make little difference.

To the extent that we can focus on our common ground within the church, we may also arrive at greater clarity in what distinguishes a Christian view from various alternatives around us, particularly naturalistic and pantheistic ones. In fact, the idiosyncratic nature of a Christian view of creation as the free work of a transcendent God may prove a surprisingly flexible and useful apologetic resource to the church in a late-modern context. This way of looking at the world has striking resonances with modern cosmological insights about the finitude of our universe; its emphasis on divine transcendence and otherness is often intriguing to secular people who sense the barrenness of naturalism; and it may open all sorts of avenues to think more freely and more constructively about aspects of human life that Christians have often underemphasized, such as (as mentioned above) the arts, vocation, and our existence as embodied creatures. Rather than directing all our energies toward the issues where Christians disagree with one another, we would do well to utilize this profound resource on which we already agree.

Second, Augustine may remind us of neglected topics within the doctrine of creation. For instance, Augustine's doctrine of creation, and in particular his robust Creator/creation distinction, can help reorient us toward the doxological and practical uses of creation. The simplicity and force of Augustine's use of the doctrine of creation in this regard is also a helpful counterbalance to the more philosophical nature of some of the previous points. In his sermons, Augustine regularly appealed to creation as an unambiguous revelation of God's goodness and wisdom. Here is a good representative example:

> Some people, in order to discover God, read books. But there is a great book: the very appearance of created things. Look above you! Look

> below you! Note it. Read it. God, whom you want to discover, never wrote that book with ink. Instead He set before your eyes the things that He had made. Can you ask for a louder voice than that? Why, heaven and earth shout to you: "God made me!"[168]

In his exegetical works as well, Augustine appeals to the pleasurable things of the world as reasons to love the God who created them. In his exposition of Psalm 104, in the midst of his allegorical use of the imagery, Augustine regularly draws observations such as the following: "Of all trees and shrubs, of all animals and flocks, and of the whole of the human race; the earth is full of the creation of God. We see, know, read, recognize, praise, and in these we preach of Him; yet we are not able to praise respecting these things, as fully as our heart doth abound with praise after the beautiful contemplation of them."[169] Similar interests in the doctrine of creation can be detected in his theoretical works. In chapter 4 we will have much to say about Augustine's interest in the goodness of tiny creatures such as insects, but here we will just give one example from *The City of God*:

> Shall I speak of the manifold and various loveliness of sky, and earth, and sea; of the plentiful supply and wonderful qualities of the light; of sun, moon, and stars; of the shade of trees; of the colors and perfume of flowers; of the multitude of birds, all differing in plumage and in song; of the variety of animals, of which the smallest in size are often the most wonderful—the works of ants and bees astonishing us more than the huge bodies of whales? Shall I speak of the sea, which itself is so grand a spectacle, when it arrays itself as it were in vestures of various colors, now running through every shade of green, and again becoming purple or blue? Is it not delightful to look at it in storm, and

[168]Sermon 126.6, *Miscellanea Augustiniana* 1:355-68, ed. G. Moran (Rome, 1930), as cited in *The Essential Augustine*, ed. and trans. Vernon Bourke (Indianapolis: Hackett, 1974), 123.

[169]Augustine, *Expositions on the Book of Psalms*, A Select Library of the Nicene and Post-Nicene Fathers of the Christian Church, First Series, ed. Philip Schaff (1888; repr., Peabody, MA: Hendrickson, 2012), 8:517.

> experience the soothing complacency which it inspires, by suggesting that we ourselves are not tossed and shipwrecked? What shall I say of the numberless kinds of food to alleviate hunger, and the variety of seasonings to stimulate appetite which are scattered everywhere by nature, and for which we are not indebted to the art of cookery? How many natural appliances are there for preserving and restoring health! How grateful is the alternation of day and night! how pleasant the breezes that cool the air! how abundant the supply of clothing furnished us by trees and animals! Who can enumerate all the blessings we enjoy?[170]

Not many theologians revel in the doctrine of creation with such energy and detail. Though Augustine's use of creation in this way is not necessarily novel, the sheer exuberance with which he pursues it may serve as a helpful reminder, stimulus, and model to evangelicals today.

Another example of how Augustine can remind us of neglected aspects of creation lies in his emphasis on the existential implications of creation. As we have seen, Augustine emphasizes the importance of the doctrine of creation for a proper understanding of human life and happiness as lived before God. Engaging Augustine's thought on this point can help us appreciate the extent to which the very concept of *creation* inherently emphasizes a vital relation to God. Creaturely existence is not an autonomous but a participatory act; it necessarily involves a dynamic orientation to God, for the only kind of being that any creature can enjoy comes from the One who is Being itself. In the *Soliloquies*, Augustine famously expressed: *noverim me, noverim te* ("I would know myself, I would know you").[171] In a sense, Augustine's entire theology is a reverberation of this plea. The relation of God and the soul is the main business of life; everything else funnels toward this. As James O'Donnell puts it, "The irrefutable solipsism of self

[170]*De civitate Dei* 22.24 (CSEL 40.2, 648).
[171]E.g., *Soliloquies* 2.1.1 (PL 32:885); cf. *De ordine* 2.18.47.

confronted with the absolute reality of God, the wholly other: all of Augustine's thought moves between those two poles."[172]

It is this interest in the God-self relation that gives Augustine's doctrine of creation such an earnest, poignant quality. Augustine conveys a sense of the reality of God invading, pressing in, touching on every level of human existence. In reading the *Confessions*, for instance, one has the sense that all of life is lived *coram Deo*, even (for instance) life as an infant.[173] Many interpreters have commented on the book's ability to communicate at this deeply interior level. Ludwig Wittgenstein regarded it as possibly the most serious book ever written.[174] Karl Jaspers said that Augustine thinks with his *blood*.[175] Miles Hollingworth describes the experience of reading Augustine as marked by

> the danger that Augustine puts you into as a reader. It is ever-present because he will always begin in some ground that you share with him and thought was safe—then he'll start pointing out the booby-traps, one by one, until the fear you taste is real, adrenal, metallic. There is a sense of relentlessness that I have not experienced with any similar writer.[176]

Few contemporary evangelicals think of the doctrine of creation as an adrenal matter; few regard Genesis 1:1-2 as a passage that sets our "heart throbbing." We need not adhere to all the nuances of Augustine's thought in order to appreciate how he can help us probe the nature of creatureliness. The doctrine of creation is more immediately concerned with shedding light on the restlessness of the human condition than with settling debates about the age of the earth.

---

[172]James J. O'Donnell, *Augustine* (Boston: Twayne, 1985), 80. Cf. his discussion of the same theme in his more recent *Augustine: A New Biography* (New York: HarperCollins, 2005), 290-91.

[173]E.g., *Confessiones* 1.6-7 (CSEL 33, 5-11).

[174]As noted in Miles Hollingworth, *Saint Augustine of Hippo: An Intellectual Biography* (Oxford: Oxford University Press, 2013), x; Fox, *Augustine*, 11, notes that Wittgenstein alluded to it fourteen times and critically engaged a quotation from it at the start of his *Philosophical Observations*.

[175]See Paul Rigby, *The Theology of Augustine's Confessions* (Cambridge: Cambridge University Press, 2015), 1.

[176]Hollingworth, *Saint Augustine of Hippo*, xi.

A third way Augustine may influence evangelical treatment of creation is by redirecting us toward new possibilities within the doctrine of creation. In other words, Augustine may not only refocus us on the center and remind us of the periphery but in some ways simply *change the subject*, introducing new categories, generating new questions and complexities, pushing the entire flow of conversation down new avenues.

For example, Augustine draws attention to the category of "good but imperfect." Evangelicals today often subtly or unconsciously assume that *unfallen* is tantamount to *perfect*, and thus disallow the category of imperfection in our view of creation. We have little or no conception of a good creation that is dynamic and developmental. Our view of pre-fallen creation is static—we think of Eden and heaven as equally immutable. Augustine also may help in this way by offering a vision of creation that is good but in development, like a lengthy poem or song. We will return to this topic in chapter 4.

A broader example concerns the significance of creation for Augustine's theology as a whole and the way it exerts a vital influence on his doctrines of sin and redemption. In much modern evangelical theology, the doctrine of creation is emphasized as a matter of apologetics, but its connection to the whole gospel is hugely underdeveloped. Sometimes it is simply floating in a vacuum, as though sin and redemption can be understood without any reference to it. In other situations, creation is the background; sin and redemption are the stage. It provides context; they fill in content.

But for Augustine, as we have seen, we cannot properly understand the nature of sin and redemption apart from creation. For instance, he maintains that creatures are radically dependent on God even apart from the contamination of sin. Thus, he regards the physical creation as mutable, inherently in need of a conversion into sharing in God's immutability (as is enjoyed by the spiritual creation) in order to fully come into itself. To be sure, sin introduces a new problem, disrupting the Creator-creature link. But in line with creation's inherently contingent status, Augustine emphasized continuity between our redemption from sin and our bodily, creaturely life.

Once again, one need not agree with the precise way Augustine goes about this in order to appreciate his ability to broaden our horizons in these areas. For instance, his emphasis on sin as privation may induce reflection on the goodness and contingency of creation as a bestowal of divine generosity and on the logical relationship of sin and its destructive consequences. Similarly, his emphasis on redemption as deification may induce reflection on the nature of redemption, and its relation to creation. Further, Augustine's affirmation of divine priority, even if we do not adhere to the precise Platonic categories in which it is cashed out, can sensitize us toward the broader issues involved in the Creator/creation distinction and trinitarian agency in the works of creation and providence. These are areas evangelicals often overlook.

In short, we must say that for Augustine, the most important aspect of the doctrine of creation is not its timing or the exact mechanics of how God does it, but rather the more basic ontological distinction it implies: that there are two kinds of reality; that the One is the source and cause of the other; and that the lesser exists in radical dependence upon the greater. God is thus the infinite reference point for all other reality: only in relation to him can anything else find "rest."

There is not a single area of theology that is unaffected by meditation on the implications of such a vision, and it is unfortunate if we pass by such considerations too quickly in our haste to determine the age of the universe.

CHAPTER TWO

THE MISSING VIRTUE IN SCIENCE-FAITH DIALOGUE

Augustine on the Importance of Humility

Let there be no obstinate wrangling, but rather diligent seeking, humble asking, persistent knocking.

De Genesi ad litteram 10.23.39

FROM HIS CONVERSION TO HIS DEATH, Augustine was captivated by Genesis 1–3. Throughout his career he tried his hand at several commentaries on these chapters, and they surface in his other works as well (as we have noticed, many have puzzled over the fact that even the *Confessions* climax into an exegesis of Genesis 1). Then, for fifteen years, he labored on a kind of *Summa Creatio*—his finished commentary on the literal meaning of Genesis (*De Genesi ad litteram*).

This body of reflection on the biblical creation and fall story is arguably unrivaled among the church fathers, yet Augustine himself was quite modest about his efforts. He depicted his first attempt at a literal commentary as oriented toward the goal of "asking questions [rather] than making affirmations."[1] Augustine would abort this

[1]*De Genesi ad litteram liber unus imperfectus* 1.1 (CSEL 28:1, 459).

project. Over twenty years later, having finally completed a literal commentary, he concludes with a disclaimer:

> I have discussed the text and written down as best I could in eleven books what seemed certain to me, and have affirmed and defended it; and about its many uncertainties I have inquired, hesitated, balanced different opinions, not to prescribe anyone what they should think about obscure points, but rather to show how we have been willing to be instructed whenever we have been in doubt as to the meaning, and to discourage the reader from the making of rash assertions where we have been unable to establish solid grounds for a definite decision.[2]

Here Augustine distinguishes the text's certainties, which ought to be affirmed and defended, from its "many uncertainties," which merit a different posture, involving inquiry (*inquirendo*), balancing (*arbitrando*), and hesitation (*ambigendo*). Augustine then derives a threefold purpose to this latter posture toward the biblical text: first, he wants to refrain from prescribing opinions in these areas; second, he wants to model teachableness; and finally, he wants to discourage his readers from making rash assertions that go beyond what we actually know. This word "rashness" (*temeritas*) will come up again and again for Augustine; we return to it below.

Later, in his *Retractations*, Augustine emphasizes the tentative nature of his finished commentary even more strongly, calling it "a work in which more questions were asked than answers were found, and of those that were found only a few were assured, while the rest were so stated as still to require further investigation."[3] Here Augustine broadens his distinction between the certain and uncertain aspects of creation to include not only relative shades of certainty, but indeed, areas of total ignorance as well. The effect of this comment is to emphasize that not only is there more ignorance than knowledge within

[2] *De Genesi ad litteram* 12.1.1 (CSEL 28:1, 379).

[3] *Retractationes* 2.50 (CSEL 36, 159-60).

the doctrine of creation, but that even most of the knowledge we have is provisional. In other words, for Augustine, we ask more about creation than we answer, and most answers generate new questions.

Those familiar with Augustine's theological stature may initially wonder whether the sense of caution reflected in such statements can be sincere. Yet anyone who takes the time to slog through Augustine's commentary work will sense a genuinely reverent, hesitating quality to them. For him, creation was a deeply mysterious doctrine, only approachable through the kind of awe that a child feels in looking up at the stars on a cloudless night. Thus, the style of his Genesis commentaries is surprisingly fluid—many passages seem to reflect an almost meandering, stream-of-consciousness style of reflection. Those who have carefully sifted through his commentary on Genesis will likely have some category for John Rist's reference to the "rambling, rhetorical feel of an Augustinian book."[4] Even when dealing with theological topics very important to him, Augustine will often assert one view, then the next page assert a different one, without any apparent concern to go back and qualify his earlier comment. His conclusions are often uncertain, or only in favor of an approximate claim. Sometimes he will launch into a topic only after the disclaimer, "Whether I am going to find and define anything certain, I do not know."[5] And all throughout his writings on creation, the tone is one of profound fascination and wonder before the mystery of God's creation; the last impression these writings give is of the crusty, hardened theologian of much popular imagination.

In this chapter we will consider how, just as Augustine's vision of creation expands our categories, his humility can inform our method.

[4]John M. Rist, *Augustine: Ancient Thought Baptized* (Cambridge: Cambridge University Press, 1994), 21.

[5]*De Genesi ad litteram* 6.29.40 (CSEL 28:1, 200). Augustine says this with respect to the creation of the soul, owning that he has not read everything written on this topic, and furthermore that some of the treatments of it "are not easy for such as I am to understand" (*De Genesi ad litteram* 6.29.40 [CSEL 28:1, 200]).

In a time when the doctrine of creation is often shrugged at, this aspect of Augustine's legacy may prove no less pertinent than the content of his views.

But this word—*humility*—provokes. We often avoid discussion of it altogether, thinking that to speak or think deliberately of humility is itself not very humble. When we do speak of humility, we typically value it as a more general virtue (useful, for instance, in conflict resolution), not specifically as a *theological* virtue (useful, for instance, in learning and speaking about God's work of creation).

Moreover, if for some reason we should be compelled to pursue humility in the context of theology, our first instinct will not likely involve looking up Augustine's name in the dictionary. Augustine is not often regarded as a particularly humble theologian. For the better part, when we think of Augustine, particularly if we have only engaged him piecemeal or indirectly, we will probably gather up in our minds some notion of an inflexible, staunch thinker—one whose deepest concerns are to define and protect the boundaries of orthodoxy, perhaps one who is reacting against his own youthful indulgences.[6]

There is, doubtless, some basis for this perception. But there is also a discernible strand in Augustine's thought, especially visible in his approach to the relation of Scripture and science, that reflects remarkable flexibility, restraint, curiosity, resilience, and—what I envision as the linchpin of all these qualities—*humility*. Humility in this context (as a theological virtue) does not entail a low opinion of oneself or one's theology, but rather a posture of eager pursuit of the truth through all the means God has provided and a ready willingness to admit what we do not yet know in the process. In his writings on creation, Augustine embodies humility in this sense.

[6]One gathers something like this picture in Peter Brown's influential biography: e.g., "A middle-aged man's sense of having once wandered, of regret at having found truth so late, will harden Augustine's attitude" (Brown, *Augustine of Hippo: A Biography*, rev. ed. [Los Angeles: University of California Press, 2000], 277).

Consider, as one introductory example, Augustine's posture toward the question of the origin of the soul in a letter to Optatus, bishop of Milevis, in fall 418:

> Before I give Your Holiness any advice on this matter, I want you to know that I never dared to state a definitive opinion on this question in my many works and never impudently dared to put into writing to instruct others what was not clear in my own mind. It would take too long to set forth in this letter the facts and reasons whose consideration has influenced me so that my assent is inclined toward neither view but remains undecided between the two. Nor is this so necessary that, without it, it is impossible to discuss this question; even if it is not sufficient for removing our hesitation, it is at least sufficient for avoiding rashness.[7]

Augustine proceeds to emphasize that the core of the Christian faith, in light of which all other matters must be regarded, is our solidarity with Adam in birth and our need for rebirth in Christ.[8] It is striking that, relatively late in his career, Augustine is content to avoid taking an opinion on such an important question. He makes it clear that his unwillingness to affirm one side or the other (that is, whether souls are propagated through natural generation, commonly called traducianism, or are directly created by God) does not stem from a lack of interest in or study of the question. Instead, he seems to regard the question as sufficiently difficult as to warrant uncertainty, such that his greatest fear is being rash.

What made Augustine more worried about rashness than hesitation? And how might his approach be instructive to us today? In what follows, we will engage Augustine to probe what humility might look like in relation to science, and then in relation to Scripture, and then in relating science and Scripture together. A working presupposition will be that

---

[7]Letter 190.2, in *Letters (Epistulae) 156–210*, ed. Boniface Ramsey, trans. Roland Teske, The Works of Saint Augustine: A Translation for the 21st Century (Hyde Park, NY: New City Press, 2004), 264.
[8]Letter 190.3, in *Letters (Epistulae) 156–210*, 265.

we do not always know in advance what humility is, and so we will conclude the chapter by considering a potential objection, gesturing toward the paradoxical nature of this particular virtue—particularly as it relates to the doctrine of creation, a topic Augustine routinely described as difficult and beyond human ability.

HUMILITY BEFORE SCIENCE: DON'T BE A "RASH, SELF-ASSURED KNOW-ALL"

In what follows we will use the term *science* in a somewhat generic and anachronistic way, referring to knowledge of the natural order arrived at through observation and experimentation—what we would categorize as natural sciences, and the ancients often conceived as a part of philosophy.[9] Augustine was interested in certain "scientific" subjects like what we would today call astronomy and geology, and in drawing up his commentary he sought to understand current opinions about topics like the cycle of the planets and the phases of the moon (while at the same time warning that the Bible did not intend to satisfy all of our curiosities on these matters).[10] Augustine uses the term *philosophi* to refer to those making claims that today would involve one or several of the natural sciences (Edmund Hill translates this term as "natural scientists" in this context).[11] Augustine also pays careful attention to what he can learn from *medici* (doctors); when discussing the body-soul relation, for instance, he pursues various theories concerning how bodily senses relate to brain activity (what today would be regarded under the label *cognitive neuroscience*).[12] At the same time, amidst our effort to detect the principles involved in Augustine's treatment of these various subjects, we

[9]A more thorough and more technically adept definition of what constitutes a "natural science" is offered by Del Ratzsch, *Science & Its Limits: The Natural Sciences in Christian Perspective* (Downers Grove, IL: IVP Academic, 2000), 13.

[10]*De Genesi ad litteram* 1.19.39 (CSEL 28:1, 28). We will discuss this passage at length below.

[11]*De Genesi ad litteram* 3.9.6 (CSEL 28:1, 71).

[12]*De Genesi ad litteram* 7.13.20 (CSEL 28:1, 212).

must of course bear in mind that the term *science* is anachronistic here and that none of these disciplines that Augustine discusses had the same role in his context as they do today.

Augustine is sensitive to the complexity of the relationship between Scripture and "science." Nowhere is this more apparent than in the beginning portions of his summative work on Genesis, the literal commentary. After an initial paragraph explaining that his work is focused on literal meaning rather than allegorical, Augustine launches into a verse-by-verse commentary starting with Genesis 1:1—but the commentary proceeds by stacking up question after question.[13] Is the "beginning" of Genesis 1:1 the beginning of time or the Son of God? Does "heaven and earth" in this verse mean the whole bodily and spiritual creation, or just bodily? When it says "and God said" throughout Genesis 1, does this refer to audible speech? How is the Trinity hinted at in Genesis 1:1-2? What is the nature of the light and darkness introduced in Genesis 1:3? What is the nature of time on the first day? Why is the sun introduced on day four after light on day one? How did morning and evening follow each other on days one to three? Where did the water that covered the earth in Genesis 1:2 go when land appeared? Where did this water (and the land it is on) come from in the first place, since it was not created on the six days? And on and on.[14]

Augustine rarely attempts to provide answers to these questions. When he addresses whether the Trinity is present in Genesis 1, a certain deliberateness and strength returns; but other than that, he often seems to be wandering. Sometimes he will list a number of interpretive possibilities but refrain from favoring one over others. On several occasions, he simply stacks up the questions without even gesturing toward possible answers.

It is striking that such a significant work—one produced over so many decades of intense intellectual struggle, and one that arguably

[13]*De Genesi ad litteram* 1.1.1-21.41 (CSEL 28:1, 1-31).
[14]*De Genesi ad litteram* 1.1.1-21.41 (CSEL 28:1, 1-31).

stands unrivaled among the church fathers—should begin in such a tentative way. Is Augustine sincerely volunteering his uncertainties, or is there a strategy to this method?

At the conclusion of book 1 we find a clue. Here Augustine anticipates the charge that his questions have not yielded answers: "Someone is going to say, 'What about you, with all this rubbing of corn in this essay, how much grain have you extracted? What have you winnowed? Why is practically everything hidden still in a heap of questions?'"[15] Augustine's sensitivity to this concern here, leading into the more natural style with which he commences in book 2 and continues throughout the rest of the commentary, suggests that his meandering style throughout book 1 is serving a purpose in the larger flow of the work. The overall rhetorical effect seems to humble the reader under the immense complexity of the subject matter. But Augustine's own immediate answer to his question also reveals the profound extent to which his method, and his purpose in the entire commentary, is developed in relation to an *apologetic* burden:

> To which I reply that I have happily reached this very food: namely that I have learned that we should not hesitate to give the answers that have to be given, in line with the faith, to people who make every effort to discredit the books our salvation depends on. So we should show that whatever they have been able to demonstrate from reliable sources about the world of nature is not contrary to our literature.[16]

What is so significant about this assertion is that Augustine not only displays interest in defending Scripture from skeptical assault, but he regards this to be the result (the "food" he has "winnowed") of all his questioning throughout this first part of the commentary. In his approach to the doctrine of creation, Augustine is not interested in merely asserting the truth of Scripture; he feels a need to harmonize

[15] *De Genesi ad litteram* 1.21.41 (CSEL 28:1, 30).
[16] *De Genesi ad litteram* 1.21.41 (CSEL 28:1, 30-31).

it with the world of nature. If the more modest approach of question-asking can better serve to answer the objections of nonbelievers, Augustine is quite happy to restrain his goals in this direction.

This concern is equally evident in the immediately prior section, where Augustine offers his most famous warning against Christian anti-scientism. This passage is frequently cited in modern creation debates, and deservedly so.[17] It is less commonly seen, however, how this passage fits into the broader context and objectives of Augustine's work. I would suggest that the whole *reason* for Augustine's questioning style throughout book 1 of his commentary is, in large part, to drive his readers to fully appreciate the concerns he articulates here:

> There is knowledge to be had, after all, about the earth, about the sky, about the other elements of this world, about the movements and revolutions or even the magnitude and distances of the constellations, about the predictable eclipses of moon and sun, about the cycles of years and seasons, about the nature of animals, fruits, stones, and everything else of this kind. And it frequently happens that even non-Christians will have knowledge of this sort in a way that they can substantiate with scientific arguments or experiments.[18]

Augustine references here a wide variety of phenomena in the natural order, ranging from the nature of rocks to the movements of constellations (what we would broadly call geology and astronomy, respectively), and then affirms the validity of knowledge about these matters

---

[17]E.g., Francis S. Collins, *The Language of God: A Scientist Presents Evidence for Belief* (New York: Free Press, 2006), 156-57 (cf. also his broader engagement with Augustine on 83, 151-53); Denis Alexander, *Creation or Evolution: Do We Have to Choose?* 2nd ed. (Grand Rapids: Monarch, 2014), 462; Denis O. Lamoureux, "Evangelicals Inheriting the Wind: The Phillip E. Johnson Phenomenon," in *Darwinism Defeated? The Johnson-Lamoureux Debate on Biological Origins* (Vancouver: Regent College Publishing, 1999), 43.

[18]*De Genesi ad litteram* 1.19.39 (CSEL 28:1, 28). What Hill translates as "with scientific arguments or experiments" here could be rendered more strictly as "by most certain reason or experiment" (*certissima ratione vel experienta*): Augustine should not be thought of as having what we mean by modern "science" exclusively in mind here, although scientific knowledge (knowledge of the material realm derived from empirical experimentation and observation) would certainly fit within the purview of his language.

among non-Christians through scientific arguments or experiments. He then continues to warn against the danger of Christians carelessly voicing foolish opinions on these topics. What is particularly striking is the severity of his rebuke:

> Now it is quite disgraceful and disastrous, something to be on one's guard against at all costs, that they should ever hear Christians spouting what they claim our Christian literature has to say on these topics, and talking such nonsense that they can scarcely contain their laughter when they see them to be *toto caelo*, as the saying goes, wide of the mark. And what is so vexing is not that misguided people should be laughed at, as that our authors should be assumed by outsiders to have held such views and, to the great detriment of those about whose salvation we are so concerned, should be written off and consigned to the waste paper basket as so many ignoramuses.[19]

This last little clause is a bit colorful in Hill's translation; "should be written off and consigned to the waste paper basket as so many ignoramuses" could be rendered more plainly as "should be blamed and rejected as so ignorant." Nonetheless, the forcefulness of Augustine's concerns is striking. The passage is so laden with hyperbole as almost to give a cartoonish impression: Christians raving (*delirare*) on in ignorance; non-Christians doubled over in irrepressible laughter, and so forth. Perhaps most striking in the Latin are the introductory terms with which Augustine condemns this kind of scenario: the words Hill translates "disgraceful and disastrous" could be rendered with more of a moral thrust as "vile" (*turpe*) and "pernicious" (*perniciosum*).

What agitates Augustine so much, as he clarifies at the end of this passage, is not that Christians are mocked, but that non-Christians, whose salvation greatly concerns him, should take the foolish opinions of some Christians as representative of the Christian faith. Assumed

---

[19]*De Genesi ad litteram* 1.19.39 (CSEL 28:1, 28-29). The phrase *toto caelo* literally means "by the whole heaven." The phrase here could be rendered "when they notice that they err as widely as possible."

in this concern is that the perception of the Christian faith as unsophisticated and ignorant is a hindrance to people's salvation—a scenario that Augustine, of course, knew all too well from experience. In the *Confessions*, in fact, he recalls how his reading "of a great many scientific books" caused him to question various claims of the Manichaean bishop Faustus.[20] As he continues in this passage of the literal commentary, Augustine extends his comments in relation to this concern:

> Whenever, you see, they catch some members of the Christian community making mistakes on a subject which they know inside out, and defending their hollow opinions on the authority of our books, on what grounds are they going to trust those books on the resurrection of the dead and the hope of eternal life and the kingdom of heaven, when they suppose they include any number of mistakes and fallacies on matters which they themselves have been able to master either by experiment or by the surest of calculations? It is impossible to say what trouble and grief such rash, self-assured know-alls (*temerarii praesumtores*) cause the more cautious and experienced brothers and sisters.[21]

Here again Augustine's concern is not simply that Christians are scientifically ignorant, but that they impute their ignorance onto authoritative Christian texts and thus impair the credibility of the Christian faith on more central matters such as resurrection, eternal life, and the kingdom of heaven. Augustine bears these concerns out in his commentaries, where he frequently leaves his interpretations of the details of Genesis 1 open-ended.

As Augustine continues beyond this often-cited passage, he explains that the rationale for his question-asking style throughout book 1 was precisely to create sympathy for this essentially apologetic concern he has just developed: "It is in order to take account of this state of things that I have, to the best of my ability, winkled out and

[20]*Confessiones* 5.3 (CSEL 33, 91); cf. *Confessiones* 5.5 (CSEL 33, 94).
[21]*De Genesi ad litteram* 1.19.39 (CSEL 28:1, 28-29).

presented a great variety of possible meanings to the words of the book of Genesis which have been darkly expressed in order to put us through our paces."[22]

Since Augustine himself recognized that the concerns evident in the famous passage of *De Genesi ad litteram* 1.19.39 were structurally and thematically determinative for the kind of commentary he aimed to write, modern readers are justified in seeing this passage as a particularly significant expression of Augustine's understanding of the relation of special and general revelation. In many respects, this passage explains why Augustine has written a commentary that contains more questions than answers. In fact, one would not be far off in conceiving Augustine's "literal commentary" as, more than a generic commentary on the text, a sustained effort at reading Genesis 1–3 in relation to apologetic concerns. There is very little in this work that does *not* reflect Augustine's effort to correlate the biblical creation account in relation to non-Christian thought.

In the sentence just cited, Augustine has stipulated that the ambiguities of Scripture have an intentional purpose: certain truths have been "darkly expressed" in order to "put us through our paces." Backing off of Hill's characteristically vivid phraseology, we might render *obscure positus* as "put obscurely" and *ad exercitationem nostrum* as "for our discipline" or "for our training." Thus, Scripture deliberately expresses some truth obscurely for our benefit. In context, the "discipline" that Augustine thinks of as resulting from the obscurity of certain biblical passages seems to involve his apologetic concern of being open to multiple interpretations of unclear passages in light of the potential conflict with scientific discovery. This, at any rate, is the concern he picks up as he continues, further justifying the style of his commentary in light of science-faith concerns: "I have avoided affirming anything hastily in a way that would rule out any

[22] *De Genesi ad litteram* 1.20.40 (CSEL 28:1, 29).

alternative explanation that may be a better one, so leaving everyone free to choose whichever they can grasp most readily in their turn, and when they cannot understand, let them give honor to God's scripture, keeping fear for themselves."[23] It is evident here that Augustine does not regard an openness to multiple possible interpretations of a particular passage to be at odds with honoring the words of Scripture—in fact, admitting uncertainty appears to be one *way* to honor Scripture.

Augustine proceeds by developing two further "scenarios" that occur in relation to the intersection of general and special revelation. If 1.19.39 of the literal commentary is concerned with ignorant Christians displaying rashness toward intelligent non-Christians, 1.20.40 concerns the danger in the opposite direction of scornful non-Christians rashly rejecting the Bible and/or upsetting the faith of weak Christians. We might chart out these scenarios in three categories:

1. Christians rashly speaking to non-Christians (1.19.39)
2. Puffed up non-Christians hastily dismissing the Bible (1.20.40)
3. Weak Christians getting intimidated by bold non-Christians (1.20.40)

Often, in contemporary retrieval of Augustine on this point, scenarios two and three are obscured while scenario one gets all the focus. It is important to consider how Augustine coaches each type of person envisioned in these scenarios, which are all too relevant in our context as well. With regard to scenario two, Augustine writes:

> But since the words of scripture that we have been dealing with can be explained along so many lines, let those people now restrain themselves, who are so puffed up with their knowledge of secular literature, that they scornfully dismiss as something crude and unrefined these texts which are all expressed in a way designed to nourish devout

[23]*De Genesi ad litteram* 1.20.40 (CSEL 28:1, 30).

> hearts. You could say they are crawling along the ground without wings, and poking fun at the nests of birds that are going to fly.[24]

Augustine bases his rebuke to this class of people on the basis of the multiple possible interpretations of Scripture ("since the words of scripture that we have been dealing with can be explained along so many lines"). Evidently Augustine regards the multifaceted nature of allowable interpretations to Scripture as a defense against the scorn of skeptics—indeed, as something that should properly humble and restrain skepticism. But Augustine also bases his rebuke here on an appeal to the *purpose* of Scripture: it is because the text of Scripture is "expressed in a way designed to nourish devout hearts" that it should not be scorned as unsophisticated. It is striking that Augustine regards the inability to appreciate the spiritual nature of scriptural language as a sufficiently serious error to warrant the metaphor he employs here: crawling animals scorning the nests of birds. Such a comparison implies that Augustine regards these proud skeptics as fundamentally failing to see what the Bible is—just as crawling beasts can only mock birds' nests if they have no conception whatever of what it is to *fly*.

Augustine continues by addressing the third scenario in science-faith dialogue:

> Some of the weaker brothers and sisters, however, are in danger of going astray more seriously when they hear these godless people holding forth expertly and fluently on the "music of the spheres" or on any questions you care to mention about the elements of this cosmos. They wilt and lose heart, putting these pundits before themselves, and while regarding them as great authorities, they turn back with a weary distaste to the books of salutary godliness, and can scarcely bring themselves to touch the volumes they should be devouring with delight.[25]

---

[24]*De Genesi ad litteram* 1.20.40 (CSEL 28:1, 30).

[25]*De Genesi ad litteram* 1.20.40 (CSEL 28:1, 30). The phrase "music of the spheres" here simply refers to the harmonious movement of the stars. See *On Genesis*, ed. John E. Rotelle, trans.

Here Augustine envisions weaker Christians taking to heart the scornful critiques of scientists and thus becoming discouraged, so that Scripture loses its delight and becomes instead distasteful to them. Augustine then develops the metaphor of extracting kernels from tough husks of wheat for how such Christians should be using the Bible against skeptical assault, concluding, "And that is why they are too lazy to use the authority they have received from the Lord to pluck the ears of wheat and go on rubbing them in their hands until they come to what they can eat."[26] That the word *laziness* would arise as a concern in this context makes clear that Augustine does not regard humility before scientific knowledge as requiring a thoughtless capitulation to whatever claims are advanced. Instead, he calls for Christians to refrain from being overawed by the claims of scientists, to exercise the proper authority God has entrusted to them, and to continue to diligently pursue the spiritual benefits of Scripture.

Augustine's concern, to be clear, is not that all Christians must learn about science. At the start of his *Enchiridion*, a handbook on the essential truths of the Christian faith, he stipulates the Christians need not live in anxiety because of their ignorance of "the properties and number of the elements, the movement and order and phases of the stars, the shape of the heavens, the kinds of animals, fruits, stones, springs, rivers, and mountains and their natures, the measurement of time and space, the indications of imminent storms and hundreds of other such things."[27] These topics Augustine associates with what the Greeks called *physikoi* (a reference to the natural philosophers of Ionia), and he stipulates that they are outside of "the sphere of religion."[28] What concerns Augustine is not that a Christian learn all of these topics, but that a

Edmund Hill, The Works of Saint Augustine: A Translation for the 21st Century (Hyde Park, NY: New City Press, 2002), 188.

[26]*De Genesi ad litteram* 1.20.40 (CSEL 28:1, 30).

[27]*Enchiridion* 3.9, in *On Christian Belief*, 277.

[28]*Enchiridion* 3.9, in *On Christian Belief*, 277.

Christian refrain from spouting off in ignorance about them. Augustine makes a similar point in the *Confessions*:

> Whenever I hear a Christian brother talk in such a way as to show that he is ignorant of these scientific matters and confuses one thing for another, I listen with patience to his theories and think it does no harm to him that he does not know the true facts about material things, provided that he holds no beliefs that are unworthy of you, O Lord, who are the Creator of them all. The danger lies in thinking that such knowledge is part and parcel of what he must believe in order to save his soul, and in presuming to make obstinate declarations about things of which he knows nothing.[29]

Here Augustine combines his concern about avoiding rash speech with his concern about associating scientific accuracy and theological knowledge. Augustine is willing to patiently tolerate Christians making scientific errors, thinking of this as essentially harmless—unless and until they claim that scientific knowledge is necessary for salvation. To do so is *presuming* and *obstinate*.

At the start of book 2 of his literal commentary, Augustine reiterates his warning against anti-scientism: "Here it occurs to me to repeat the warning I gave in Book I about the mistake of relying on the evidence of a scriptural text against those who produce these subtle arguments about the weights of the elements."[30] He quotes Psalm 136:6 ("who founded the earth on the water") as one possible passage that can be mentioned in this context. Augustine's concern is that those making the argument from the weights of the elements "do not acknowledge the authority of our literature and are ignorant of the way in which that was said, and so they are more likely to poke fun at the sacred books than to repudiate what they have come to hold by reasoned arguments or have proved by the clearest experiments."[31]

[29]*Confessiones* 5.5 (CSEL 33, 95).
[30]*De Genesi ad litteram* 2.1.4 (CSEL 28:1, 34).
[31]*De Genesi ad litteram* 2.1.4 (CSEL 28:1, 34).

Part of Augustine's concern here is that non-Christians are ignorant of how Scripture communicates. That is why they ridicule it. But Augustine is also concerned that Christians reinforce this misperception when they handle Scripture too woodenly. He suggests that Psalm 136:6, for instance, can be taken figuratively, with "heaven" and "earth" signifying the "spiritual and fleshly-minded members of the Church respectively."[32] He further stipulates that if we are obliged to take the verse literally, the reference to the earth here can be interpreted not unreasonably as referring to the "heights of the earth" above the water or the roofs of caverns overhanging the earth.[33]

What is especially noteworthy is the dexterity with which Augustine is willing to approach interpreting Scripture in the face of an apparent conflict with genuine knowledge of the natural order. At no point in this context does Augustine suggest that a scientific claim be simply rejected because it is at odds with Scripture. Rather, he poses several different ways of reconsidering a particular interpretation in order to harmonize it with scientific knowledge. In summarizing Augustine's thought on the relation of science and faith, Peter Harrison calls this principle of Augustine's thought "The Priority of Demonstration," which he defines as: "When a conflict appears between a proven truth about nature and an interpretation of scripture, scripture should be reinterpreted."[34]

At the same time, Augustine is perfectly willing, in the appropriate context, to wield biblical authority to blunt the force of a skeptical claim. For instance, just a short while later, he responds to the claim that the coldness of the planet Saturn prohibits any water from being there by insisting: "In whatever form, however, waters may be there, and of whatever kind, let us have no doubts at all that that is where they

[32]*De Genesi ad litteram* 2.1.4 (CSEL 28:1, 34).

[33]*De Genesi ad litteram* 2.1.4 (CSEL 28:1, 34-35).

[34]Peter Harrison, "Is Science-Religion Conflict Always a Bad Thing? Augustinian Reflections on Christianity and Evolution," in *Evolution and the Fall*, ed. William T. Cavanaugh and James K. A. Smith (Grand Rapids: Eerdmans, 2017), 213.

are; the authority of this text of scripture, surely, overrides anything that human ingenuity is capable of thinking up."[35] Similarly, Augustine will later assert that when a biblical claim strikes us as silly but it entails no contradiction, we must trust that it was stated the way it was "so that it might be entrusted to our hearts, given the supremely trustworthy authority of the holy scriptures, as something mystical."[36] Thus, it is not Augustine's position that, whenever the Bible says something difficult or embarrassing, we must move laterally around it somehow. Augustine is content to say with Paul, "Let God be true though every one were a liar" (Rom 3:4). Harrison defines this principle of Augustine's thought "The Priority of Scripture," which he defines as: "When there is an apparent conflict between scripture and doctrine about the natural world based on reason or sense, where the latter doctrine is not demonstrated, the literal reading of scripture should prevail."[37]

In considering these two different aspects of Augustine's thought, it is evident that he is not generically skeptical of science. What is most striking, perhaps, is how high his view of science is: he only opposes a particular scientific claim on a biblical basis when the science in question is not established, and overall his greater emphasis falls on how commonly a conflict between science and Scripture results from a mistaken interpretation of the Bible. Later, for instance, when dealing with the phrase "who stretched out the sky like a skin" (Ps 104:2) and its apparent contradiction of those who claim the sky is a sphere or a globe, Augustine again affirms the superiority of the divine authority of Scripture over the "guesswork of human weakness."[38] At the same time, he makes clear in the way he continues that this appeal to biblical authority should only be made against weaker kinds of knowledge about the natural order: "But if it should happen that they can prove their case with evidence and arguments

[35]*De Genesi ad litteram* 2.5.9 (CSEL 28:1, 39).
[36]*De Genesi ad litteram* 9.12.22 (CSEL 28:1, 283).
[37]Harrison, "Science-Religion Conflict," 213.
[38]*De Genesi ad litteram* 2.9.21 (CSEL 28:1, 46).

beyond any possibility of doubt, then it has to be demonstrated that what is said here among us about a skin is not contrary to those explanations of theirs."[39] He then appeals to the assertion in Isaiah 40:22 that the sky is suspended like a dome, warning that a crassly literal understanding of Psalm 104:2 will lead to conflict with this passage as well. He then reasons:

> But if we are obliged, as indeed we are, to understand these two expressions in such a way that they are found to agree (*concordare*) with each other and not to be in the least contradictory, then we are also and equally obliged to demonstrate that neither of them is opposed to those explanations, should they happen to be shown by rational arguments to be true, which inform us that the sky has the shape of a hollow globe all round us—provided, once again, it can be proved.[40]

In this context, Augustine's concern for scriptural statements to be harmonized with science appears to have the same forcefulness as his concern that scriptural statements be harmonized with each other. If Isaiah 40:22 and Psalm 104:2 must have concord with each other, so must both verses with science (again, assuming the science is legitimate). This way of reasoning reveals a sincere respect for genuine discoveries of science. Augustine proceeds to assert that both *dome* and *skin* can be taken figuratively, while also exploring other ways to reconcile them "to satisfy the tiresome people who persist in demanding a literal explanation."[41] As Edmund Hill notes, to find literalism "tiresome" while writing a "literal" commentary reveals much about Augustine's conception of literal meaning—a theme we will pick up in the next chapter.[42]

---

[39]*De Genesi ad litteram* 2.9.21 (CSEL 28:1, 46).

[40]*De Genesi ad litteram* 2.9.21 (CSEL 28:1, 46-47).

[41]*De Genesi ad litteram* 2.9.22 (CSEL 28:1, 47).

[42]Hill suggests that this reference is "very revealing about his real intentions, and his thoroughly 'anti-fundamentalist' understanding of the literal sense" (Augustine, *On Genesis*, ed. John E. Rotelle, trans. Edmund Hill, The Works of Saint Augustine: A Translation for the 21st Century [Hyde Park, NY: New City Press, 2002], 202).

Part of Augustine's reluctance to remain unflinching in a particular biblical interpretation in the face of scientific pressure stems from his emphasis on the spiritual purpose of Scripture. The whole issue of the shape of the sky, he insists, is a topic that "our authors with greater good sense passed over as not holding out the promise of any benefit to those wishing to learn about the blessed life, and, what is worse, as taking up precious time that should be spent on more salutary matters."[43] He then admits:

> But because the trustworthiness of the scriptures is here in question, this, as I have reminded readers more than once, has to be defended from those who do not understand the style of the divine utterances, and who assume when they find anything on these matters in our books, or hear them read out from them, which seems to be contrary to explanations they have worked out, that they should not place any confidence in the scriptures, when they foretell or warn or tell them about other useful things.[44]

In other words, Augustine thinks the whole issue of how the sky is shaped is unimportant to the Bible's spiritual purpose, but needs to be addressed anyway because for some people its overall effect is to undermine confidence in the Bible. He does not believe that the biblical authors are necessarily ignorant about scientific matters, but he supposes that they largely passed over these matters in order to focus on their essentially spiritual purpose. Thus, he concludes, "Our authors knew about the shape of the sky whatever may be the truth of the matter. But the Spirit of God who was speaking through them did not wish to teach people about such things which would contribute nothing to their salvation."[45] All this suggests a deep wariness in Augustine about pressing the Bible into the service of concerns that are alien to its own.

[43] *De Genesi ad litteram* 2.9.20 (CSEL 28:1, 45).
[44] *De Genesi ad litteram* 2.9.20 (CSEL 28:1, 45-46).
[45] *De Genesi ad litteram* 2.9.20 (CSEL 28:1, 46).

Augustine is also leery about simple appeals to divine omnipotence to resolve science-faith tensions. At the start of book 2 of his literal commentary, in dealing with the "expanse" or "firmament" referenced in Genesis 1:6-8, he shows awareness that many people insist that water cannot exist above the heavenly constellations because gravity keeps water on earth, even if as a vapor. He is keen that this concern not be naively and quickly sloughed off:

> Nor should anybody try to refute them by appealing to the omnipotence of God, *for whom all things are possible* (Mk 10:27), and saying we just have to believe that he can cause even water, as heavy as what we know by our own experience, to spread out over the substance of the heaven or sky in which the stars have their place.[46]

Augustine dislikes this overly quick appeal to God's power because, he reminds his readers, the aim of his commentary is to inquire after what Scripture teaches about God's work of creation, not what we might *wish* to work out of Scripture.[47] Thus, rather than simply asserting that God can do anything, Augustine is willing to entertain "how?" questions—not because he doubts Scripture's integrity, but because he wants to be sure he has not misunderstood its meaning.

It would be misleading to think that Augustine has no worries about the misuse of science. He is very firm in his denunciation of certain kinds of study of the natural order, such as astrology and divination, which he regards as often facilitated by consorting with demons.[48] He is also aware that a little scientific knowledge is often more dangerous than none. For instance, in commenting on the nature of God's rest on day seven of Genesis 1, he begins to speculate about the perfection of the number seven (as he had done with the number six), but then hesitates: "I fear lest I should seem to catch at an opportunity of airing my little smattering of science more childishly than profitably."[49]

---

[46]*De Genesi ad litteram* 2.1.2 (CSEL 28:1, 32-33).
[47]*De Genesi ad litteram* 2.1.2 (CSEL 28:1, 33).
[48]*De Genesi ad litteram* 2.17.35-2.17.37 (CSEL 28:1, 59-62).
[49]*De civitate Dei* 11.31 (CSEL 40.1, 559).

Overall, Augustine does not advocate for an unrestrained subservience to scientific claims. At the same time, his respect for the validity of well-founded scientific knowledge is profound, and he exerts extreme care at attempting to maintain harmony between this knowledge and the claims of Scripture. While he maintains the Bible's infallible trustworthiness, his emphasis more often falls on our fallibility as its interpreters, and thus our need for caution in the face of a scientific challenge.

HUMILITY BEFORE SCRIPTURE: DON'T BE "HEAD OVER HEELS IN HEADSTRONG ASSERTION"

Augustine often appears to be struggling in his commentaries. This very point may be instructive, for some modern biblical commentaries do not easily give the impression that uncertainty is permissible. Sometimes modern exegetes, in fact, seem to blithely glide by interpretations over which the great bishop of Hippo agonized, offering confident answers where he was fearfully silent, or assuming the simplicity of a conclusion that he has advanced only with painstakingly nuanced caveats and clarifications. Even where we might differ with Augustine's conclusions, we might still benefit from considering the toiling nature of his method. It is difficult to labor alongside Augustine in Genesis 1–3 and then require modern readers to regard its interpretation as self-evidently clear.

To single out a term mentioned before: Augustine is always warning against the danger of "rashness" (Latin: *temeritas*). This word comes up again and again in his commentaries. One of his signature maneuvers is to canvass a number of interpretive options, gesture toward a possible answer, but ultimately refrain from requiring a definitive stance from his reader. For instance, he offers two possible ways to understand the "expanse" of Genesis 1:7, and then counsels, "You may choose whichever you prefer; only avoid asserting anything rashly, and something you don't know as if you did; and remember you are just a human being investigating the works of God to the extent you are permitted to do so."[50]

Often, after surveying the details of possible interpretations, Augustine will simply double back to emphasize the main point of a passage. For example, after outlining several ways of understanding the Spirit's hovering over the waters in Genesis 1:2, he ultimately falls back on a more basic appeal: "But whichever of these opinions is true, we are bound to believe that God is the author and founder of all things that have originated, both those that are seen and those that are not seen."[51] Augustine frequently displays a willingness to simply suspend judgment about a particular matter of interpretation until more information comes in. At one point in his discussion of the nature of the body of Adam, for instance, he states that "we are still in no inordinate hurry to affirm anything for sure, but instead we are waiting to see whether the rest of scripture does block this understanding of things."[52]

For Augustine, humility before Scripture entailed this kind of willingness to countenance multiple interpretations of unclear passages. But it also entailed a restraint in how we wield even those passages that we think allow only one interpretation, and a willingness to pursue that meaning patiently, with a scrutinizing attention to our own fallibility:

> In discussing obscure matters that are far removed from our eyes and our experience, which are patient of various explanations that do not contradict the faith we are imbued with, let us never, if we read anything on them in the divine scriptures, throw ourselves head over heels into the headstrong assertion of any of them. Perhaps the truth, emerging from a more thorough discussion of the point, may definitively overturn that opinion, and then we will find ourselves

---

[50]*De Genesi ad litteram liber unus imperfectus* 9.30 (CSEL 28:1, 481).
[51]*De Genesi ad litteram liber unus imperfectus* 4.18 (CSEL 28:1, 470-71).
[52]*De Genesi ad litteram* 6.28.39 (CSEL 28:1, 200).

> overthrown, championing what is not the cause of the divine scriptures but our own, in such a way that we want it to be that of the scriptures, when we should rather be wanting the cause of the scriptures to be our own.[53]

Augustine is shrewdly sensitive here to how easy it is to say we are defending the Bible when we are really defending ourselves. He urges caution in arriving at an interpretation of Scripture; since we often arrive at the truth only after a thorough examination of a point, a hastily drawn conclusion is liable to reflect our own interests rather than those of Scripture. Augustine also emphasizes here the *difficulty* of the doctrine of creation, since it is a matter removed from our ordinary observation and experience, and it concerns topics that are "patient of various explanations," all within the rule of faith.

To be clear, Augustine also regards attributing error to Scripture as inexcusable rashness. In the *Confessions* he urges Christians to make every effort to understand a scriptural text, since its author wrote only the truth, and "we are not so rash as to suppose that he wrote anything which we know or think to be false."[54] Similarly, when in the literal commentary Augustine considers the possibility of science overturning an interpretation of Scripture, he makes it clear that "if this does happen, then this is not what divine scripture contained, but what human ignorance had opined."[55]

Nonetheless, those who uphold an infallible text must always remember that they only read and use it fallibly. This is a hallmark of Augustine's approach to Scripture: his dual emphasis on both the infallibility of the text and the fallibility of its interpreters. For instance, in book 12 of the *Confessions*, after listing five possible interpretations of Genesis 1:1 and then five possible interpretations of Genesis 1:2, he distinguishes two kinds of disagreements Christians can have: "We

[53]*De Genesi ad litteram* 1.18.37 (CSEL 28:1, 27).
[54]*Confessiones* 12.18 (CSEL 33, 328).
[55]*De Genesi ad litteram* 1.19.38 (CSEL 28:1, 28).

may disagree either as to the truth of the message itself or as to the messenger's meaning."[56] He then stipulates that while he wants nothing to do with those who deny the truthfulness of the biblical text, he nonetheless desires unity, joy, charity, and mutual learning among those who disagree on its meaning: "I wish to have (no dealing) with any who think that Moses wrote what was not true. But I pray that in you, O Lord, I may dwell in harmony and joy with those who feed upon your truth in the fullness of charity. May they and I together approach the words of your Book, and in them may we seek your meaning as we were meant to understand it."[57]

Augustine proceeds to affirm an unshakable confidence that Moses wrote the truth in Genesis 1:1, yet confesses that he does not have "equal confidence" regarding differences in how to render this verse.[58] He then develops his earlier appeal for charitable disagreement among Christians by envisioning a scenario in which another Christian acknowledges that Augustine's interpretation is not inconsistent with the truth, and yet insists that "Moses did not mean what you say. He meant what I say."[59] In such a scenario, Augustine asks God to help him not be impatiently irritated: "I beg you to water my heart with the rain of forbearance, so that I may bear with such people in patience."[60] In fact, strikingly, Augustine attributes the insistence on one's own opinion in such a situation to stubbornness and pride: "They are in love with their own opinions, not because they are true, but because they are their own."[61]

Augustine does not suffer from a lack of eagerness to pursue Scripture's meaning. He himself labors on and on to find it. At the same time, he emphasizes the need to patiently endure different (orthodox)

---

[56]*Confessiones* 12.23 (CSEL 33, 333).
[57]*Confessiones* 12.23 (CSEL 33, 333).
[58]*Confessiones* 12.24 (CSEL 33, 333-34).
[59]*Confessiones* 12.25 (CSEL 33, 334).
[60]*Confessiones* 12.25 (CSEL 33, 335).
[61]*Confessiones* 12.25 (CSEL 33, 335).

views. Against those who insist on their own construal of Moses' meaning, his language is strong: "It appalls me, because even if their explanation is the right one, the arbitrary assurance with which they insist upon it springs from presumption, not from knowledge. It is the child of arrogance, not of true vision."[62] Augustine reflects a concern here that having a right interpretation is not all that matters; how we arrive at and maintain that interpretation is also important. If we have the right view because of an arbitrary presumption, it is not the truth that compels us, but pride. Augustine then ties this presumption to the mistaken idea that any one Christian possesses God's truth, which is in fact the property of all Christians: "Your truth is not mine alone nor does it belong to this man or that. It belongs to us all. . . . If any man claims as his own what you give to all to enjoy and tries to keep for himself what belongs to all, he is driven to take refuge in his own resources instead of in what is common to all."[63] Augustine seems to regard the interpretation of Scripture in far more communal terms than many modern Christians: to rely upon our own private judgments, irrespective of what other Christians have thought, is a form of "taking refuge in our own resources." Augustine evidently believes that God gives the truth of Scripture to all Christians, and thus we need each other to understand it. The logic, roughly, seems to be that because God gives Scripture to all Christians, he therefore gives its interpretation to the same.

Augustine then offers a solemn warning: "When so many meanings, all of them acceptable as true, can be extracted from the words that Moses wrote, do you not see how foolish it is to make a bold assertion that one in particular is the one he had in mind?"[64] The strength of this appeal, as Augustine continues, seems to stem from his belief that this way of handling Scripture actually runs against the whole purpose

[62]*Confessiones* 12.25 (CSEL 33, 335).
[63]*Confessiones* 12.25 (CSEL 33, 335).
[64]*Confessiones* 12.25 (CSEL 33, 336).

for which Scripture was written in the first place: "Do you not see how foolish it is to enter into mischievous arguments which are an offence against that very charity for the sake of which [Moses] wrote every one of the words that we are trying to explain?"[65] Thus, in Augustine's mind, arrogant argumentativeness about the meaning of Scripture is out of accord with the whole aim of the meaning that is being disputed: namely, that it would produce charity.

Humility before Scripture also entailed for Augustine a willingness to advance certain interpretations on a provisional basis, as something other than the final word on a passage. After developing his view that the days of Genesis 1 are described with reference to modes of angelic knowledge, a topic we will return to in the next chapter, Augustine makes a significant concession:

> So then, if anybody is not satisfied with the line which I have been able in my small measure to explore or to trace, but requires another theory about the numbering of those days, by which they may be better understood, not as prophetic types and figures, but as a strict and proper account of the way foundations of this creation were laid, then by all means let him look for one and with God's help find one. I am certainly not insisting on this one in such a way as to contend that nothing else preferable can be found, while I do insist that holy scripture did not intend to suggest to us that God rested as if he were tired out, or racked with anxiety and worry.[66]

Evidently, Augustine regards the precise nature of the days of Genesis 1 as more in the realm of the "uncertainties" to be treated carefully, not the "certainties" to be defended staunchly (more on this in the next chapter). Of course, even here Augustine's flexibility has boundaries; he is not willing to countenance a weary deity. But outside the rule of faith, his openness to alternative views is remarkable—not least because he has spent so much time laboring on his own.

---

[65]*Confessiones* 12.25 (CSEL 33, 336).

[66]*De Genesi ad litteram* 4.28.45 (CSEL 28:1, 127).

Similarly, Augustine holds that our engagement with Scripture on difficult topics will sometimes yield only varying and partial approximations of the truth. Some aspects of the doctrine of creation, for instance, Augustine regards as "so far removed from our senses and from what ordinary human thought is used to, that we have first of all to believe them on divine authority, and then to come to some kind of knowledge of them from the things we already know, in the greater or lesser degree of which we are capable with divine assistance."[67] Thus, even assuming both the acceptance of the divine authority of the Scripture and a reliance on God's help, we will inevitably progress in our understanding of some topics better than we do in other topics.

This willingness to wait for further insights, and proceed slowly and progressively toward the truth in our understanding—what we might call *hermeneutical patience*—is an important feature of Augustine's approach to Scripture. Thus, in his literal commentary Augustine commends the "restraint which requires us, when faced with the profundities of Scripture, to be painstaking in our researches rather than cavalier (*temeritatem*) in our assertions."[68] Later, at the beginning of book 7, where he takes up a challenging question concerning the origin of the soul, he affirms his desire to read Scripture correctly. His stated definition of reading Scripture this way, intriguingly, involves the process as well as the outcome of interpretation:

> "Correctly," though, means truly and relevantly, without either brashly refuting or brazenly asserting anything while there is still any doubt over its being true or false, whether in light of the Christian faith or of Christian learning; but unhesitatingly affirming whatever can be taught on the clear evidence of facts and by the light of reason, or on the unambiguous authority of the scriptures.[69]

[67] *De Genesi ad litteram* 5.12.28 (CSEL 28:1, 156).
[68] *De Genesi ad litteram* 6.9.14 (CSEL 28:1, 180).
[69] *De Genesi ad litteram* 7.1.1 (CSEL 28:1, 201).

Augustine's characteristic concern about "rashness" is evident here in his assertion that a correct interpretation of Scripture involves not merely faithful adherence to what it positively asserts, but also a prudent silence over those areas in which the truth is not yet clear. His distinction between "the Christian faith" and "Christian learning," combined with his reference to "clear evidence of facts and . . . the light of reason" in addition to the authority of Scripture, seems to suggest that Augustine does not conceive of a narrow biblicism (in which we remain within the confines of explicit biblical assertions) as reading the Bible "correctly." Rather, he calls for an eager pursuit of Scripture's meaning through every available avenue of knowledge, even while he simultaneously opposes anyone "brashly refuting" that meaning.

Augustine pursues this topic through book 7, exploring all kinds of questions pertaining to the origins and nature of the soul with an evident open-mindedness and curiosity—but ultimately concluding by identifying a few essential truths about the soul that must be maintained (it comes from God, it is not of God's substance, it is incorporeal, etc.).[70] What Augustine then identifies as the overriding *value* of his inquiry, however, is not this conclusion, but rather the tentative, inquiring method he has used throughout to arrive at it:

> I trust that everything else I have said in this volume by way of discussion will at least have this value for readers, that they will either have learned how matters on which scripture does not speak plainly are to be investigated without making rash assertions or else, if they are not satisfied with this manner of investigation, that they will let me know how they would set about it. Then if they have something to teach me, I hope I will not reject it, while if they have not, let them join me in seeking the one from whom we all have lessons to learn.[71]

---

[70]*De Genesi ad litteram* 7.28.43 (CSEL 28:1, 228).
[71]*De Genesi ad litteram* 7.28.43 (CSEL 28:1, 228).

Perhaps nowhere else in all his commentary is the posture of *learning* that marks Augustine's method so potent. He not only offers his characteristic warning about rash assertions, but he even backs off this claim, inviting alternative approaches from his readers. There is no reason to attribute insincerity to his expression; coming at the end of such a tentative book, in which he has been genuinely struggling toward an answer, his comments here have the ring of authenticity. It is evident that Augustine regards the pursuit of truth as more valuable than any of his own particular efforts at finding it—hence his willingness to lay aside his own view and learn from his readers.

It is clear that hermeneutical patience in no way implied for Augustine a carefree apathy in his exegesis; he was painstaking in his efforts to progress as far as he could in his understanding of Scripture. In discussing the origins of the soul in book 10 of his literal commentary, Augustine canvasses three possible views and states his intention to eagerly explore the truth, with God's help. He expresses his hope of arriving at a view that can at least provisionally be held with confidence: "Let us see if we cannot perhaps come to a conclusion of such general acceptability that it would not be absurd for us to hold it until something more certain emerges."[72] Nonetheless, Augustine recognizes that he may not arrive at an answer at all—but he conceives of *that very uncertainty* as a valuable outcome: "If we cannot even manage this, with the weight of the evidence wobbling equally now on this side of the scales, now on that, it will at least be clear that our doubts and hesitations are the result not of our shirking the labor of investigation, but of our shunning the rashness of dogmatic assertion."[73] This passage highlights Augustine's profound appreciation of the struggle that can attend interpreting difficult passages, as well as his intuition that the ultimate goal of this struggle is something

[72]*De Genesi ad litteram* 10.3.6 (CSEL 28:1, 300).
[73]*De Genesi ad litteram* 10.3.6 (CSEL 28:1, 300).

more complex than simply arriving at a right answer. For Augustine, our very ignorance may convey the larger benefit of "shunning the rashness of dogmatic assertion."

Augustine can be open to uncertainty because he regards the purpose of theological inquiry to be godliness, and he is willing to leave questions unanswered to the extent that they do not lead to this end: "It should be enough for our faith that we should know where we are going to arrive by living devout and loyal lives, even if we are ignorant of where we have come from."[74] He furthermore assumes that we do not always know in advance what will lead to godliness, and so there should be an openness and humility in the posture with which we inquire about the doctrine of creation: "At least let there be no obstinate wrangling, but rather diligent seeking, humble asking, persistent knocking, so that if it is expedient for us to know this, the one who certainly knows better than we do what is expedient for us may also give us this, the one who gives good gifts to his children."[75] One of the reasons, in fact, that Augustine is patient of having multiple interpretations of difficult passages is his concern for the spiritual *consequences* of particular interpretations. Thus, in the *Confessions* he asks, "How can it harm me that it should be possible to interpret these words in several ways, all of which may yet prove to be true? How can it harm me if I understand the writer's meaning in a different sense from that in which another understands it?"[76] In this context, Augustine emphasizes that certain kinds of errors—for instance, those that stray from what the biblical author had in mind, but nonetheless are true to the rule of faith as a whole—are not likely to produce spiritual damage, and must thus be treated differently than errors that imperil the faith of those advancing them.

[74] *De Genesi ad litteram* 10.23.39 (CSEL 28:1, 326).
[75] *De Genesi ad litteram* 10.23.39 (CSEL 28:1, 326).
[76] *Confessiones* 12.18 (CSEL 33, 328).

HUMILITY IN RELATING SCRIPTURE AND SCIENCE

This brief overview is likely to have already provoked some critical reflection on current evangelical practices and modes of thought. We now pause to recognize and extend several particular insights, especially with a view to the *relationship* between science and Scripture. If evangelicals were to consider Augustine's model of humility before both Scripture and science, how might this help us navigate the tensions currently bubbling up between these two domains? We cannot be exhaustive here, so we will offer three observations, each in relation to contemporary views, assumptions, or methods.

The first and perhaps most obvious point is that Augustine's example will not encourage a simple and unconsidered rejection of any scientific claim that appears to be in conflict with traditional theology or interpretations of Scripture. We have observed Augustine's curiosity regarding the topics pursued by *philosophi* and *medici*, and his respect for the knowledge acquired by observation and reason. We have watched how eagerly Augustine attempts to harmonize biblical statements with the claims of science, how flexible he is in his interpretations, how rarely he simply rejects the science out of hand. We have felt the sting of his censure against the complacent anti-scientism and anti-intellectualism that Christians sometimes exhibit (which Augustine calls "disgraceful and disastrous")—and how this concern is structurally and thematically significant for his crowning work on Genesis. We have bumped up against, again and again, his warning against *temeritas*, and his apparent fear of being rash much more than being irresolute and vacillating. We have noticed the carefulness and self-awareness with which he navigates his distinction between "opinions" outside the rule of faith and "certainties" within it. We have witnessed his hermeneutical patience in enduring various interpretations, in holding some views provisionally, in altogether refraining from judgment about unclear matters, and in warning us against the pride of stubbornly clinging to isolated interpretations. We have perceived the vexed, almost tortuous painstakingness with which he advocates for his own views: for example, how he will own his lack of reading and lack of understanding of a topic;[77] how he can invite his readers to pursue a different view from his own;[78] how he wants to learn from his readers "if they have something to teach me;"[79] and on and on we might go.

One does not need to be particularly well-networked or well-traveled to recognize that contemporary evangelical treatments of the doctrine of creation often do not share these values. Indeed, most of those operating in evangelical contexts have observed approaches to creation that are characterized (to use Augustine's terms) by more "obstinate wrangling" and less "diligent seeking, humble asking, persistent knocking." This is particularly evident in our attitude toward the natural sciences. In some evangelical circles, science is viewed with suspicion, associated with secularism and naturalism, and held at arm's length from any meaningful discourse with theology. We occasionally hear the assertion that science should never change our theology or our interpretation of the Bible, and some voices give the impression that any such adjustments are the result of worldliness or the "fear of man."

Of course, this streak of anti-scientism is not universal in evangelicalism. Huge numbers of evangelicals, especially younger ones, are actively wrestling with and engaging the challenge of relating the claims of twenty-first-century science to the Christian faith. Many prominent Christians are more open to evolution, for instance.[80] In some circles, the temptation may be to be *overly* deferential to science.

---

[77]*De Genesi ad litteram* 6.29.40 (CSEL 28:1, 200).

[78]*De Genesi ad litteram* 4.28.45 (CSEL 28:1, 127).

[79]*De Genesi ad litteram* 7.28.43 (CSEL 28:1, 228).

[80]A number of stories in this regard are recounted in *How I Changed My Mind About Evolution: Evangelicals Reflect on Faith and Science*, ed. Kathryn Applegate and J. B. Stump (Downers Grove, IL: IVP Academic, 2016).

At the other end of the spectrum, enormous energy is expended toward developing, promoting, and defending alternative models to mainstream science (often called "creation science")—particularly since the publication of *The Genesis Flood* in 1961 by John C. Whitcomb and Henry M. Morris.[81] Although the view popularized within this movement—often called "young-earth creationism"—was not the stock response of conservative Christians to the discovery of testimonies of age in the universe starting in the late eighteenth and early nineteenth centuries, it has become common in many conservative evangelical circles today.[82] One could make the argument that the "science" displayed in this movement is sufficiently revisionist (despite claims and efforts to the contrary[83]) that it still merits Augustine's rebukes.[84]

But here I am thinking more of the general posture of apathy toward the natural sciences that often characterizes more conservative and moderate evangelicals between these two camps. To make a generalization, there is a greater tendency toward a default distrust and/or dismissiveness toward scientific claims within contemporary evangelical Christianity, especially in the United States, than has historically been the case throughout catholic, historic Christianity.[85] Augustine is a mirror that reflects this fact. In contrast to his hesitations, evangelical appeals to peace and moderation are rebuked as

[81]John C. Whitcomb and Henry M. Morris, *The Genesis Flood: The Biblical Record and its Scientific Implications* (Philadelphia: Presbyterian and Reformed, 1961).

[82]We will document this claim in chapter 5.

[83]E.g., *In Six Days: Why Fifty Scientists Choose to Believe in Creation*, ed. John Ashton (Green Forest, AR: Master Books, 2000).

[84]Even the so-called father of young-earth creationism, Henry Morris, could make claims such as, "If the Bible and Christianity are true at all, the geological ages must be rejected altogether" (Henry M. Morris, *Scientific Creationism* [Green Forest, AR: Master Books, 1985], 255).

[85]In addition to engagement with Augustine as an example of this point, we might think of the founding of modern science itself, many of the pioneers of which were devout Christians, such as Francis Bacon, Isaac Newton, Johannes Kepler, Robert Boyle, Galileo Galilei, Michael Faraday, Louis Pasteur, and Blaise Pascal. It is fascinating to read Boyle's lectures, for example, starting in the 1690s, and see the interplay of biblical exposition with scientific theory in many of the early ones. It paints quite a different picture than the warfare model of science and faith held by many today.

"trembling at the words of scientists";[86] appeals to specific points of data "require no reply" because "we simply have no biblical information on this subject";[87] and attempts to harmonize Scripture and science are seen as compromise: "A choice has to be made between Scripture, which is authored by God, and modern science, which is authored by men."[88] I can recall many conversations about creation in evangelical circles in which reference to scientific evidence is met with indifference ("Who cares about that?"), surprise ("What value could that have?"), and/or suspicion ("Why are you studying that?").

Augustine's approach, by contrast, reminds us that it is not necessarily a mark of convictional weakness or doctrinal erosion to listen carefully to scientific claims, or to consider if we may have misinterpreted a given biblical text, or to be open to theological reformulation when necessary. Now, of course, listening to the science will not always entail that we will agree with its claims. But Augustine would encourage us to "do our homework" before arriving at that conclusion, and he would caution us against too quickly assuming the immovability of an established interpretation. Much truth comes to us in the categories of common grace and natural revelation, and Augustine was willing to test and reconsider his views against advances in these areas. He was also sensitive to the damage that can ensue when we cling to erroneous interpretations against good reason and evidence.

I suggest that the church of Jesus Christ, as we grapple with the array of scientific challenges and questions that face us in the early twenty-first century, should heed Augustine's appeal to pursue "diligent

---

[86]Theodore J. Cabal and Peter J. Rasor, *Controversy of the Ages: Why Christians Should Not Divide Over the Age of the Earth* (Wooster, OH: Weaver, 2017), 150; this is part of Terry Mortenson's response to Cabal's 2001 Evangelical Theological Society presentation.

[87]J. Ligon Duncan III and David W. Hall, "The 24-Hour View," in David G. Hagopian, ed., *The Genesis Debate: Three Views on the Days of Creation* (Mission Viejo, CA: Cruxpress, 2001), 111. Earlier, they claim that even the best of "secular science" is "only tentative, changing" (57), and describe evidences for the age of the universe as "passing scientific trends" (61).

[88]Andrew A. Snelling, *Earth's Catastrophic Past: Geology, Creation & the Flood*, vol. 1 (Dallas: Institute for Christian Research, 2009), 5.

seeking, humble asking, persistent knocking" rather than "obstinate wrangling," and to be "painstaking in our researches rather than cavalier in our assertions."[89] Augustine's legacy prods us toward these values. He reminds us that avoiding rashness is sometimes more important than avoiding hesitation, that our posture matters as well as our position, and that the virtue of courage is best expressed in conjunction with the corresponding virtue of humility (and is often dangerous without it).

Second, though, in the other direction, we must also consider Augustine's unflinching commitment to the truth of Scripture in the face of apparent challenges and areas of tension. If Augustine's respect for science made him willing to reconsider his interpretations, his reverence before Scripture made him unwilling to ever question or doubt the text itself. Thus, Augustine was willing to resist scientific claims, and he emphasized the need to trust the Bible even when it is difficult to understand. In fact, in his commentaries Augustine is happy, when appropriate, to set the "supremely trustworthy authority of the holy scriptures"[90] over and against the "guesswork of human weakness."[91]

Augustine is particularly burdened that the attempt to relate biblical claims to the natural world in no way impugns the integrity and trustworthiness of Scripture. For instance, in the context of claiming that it is permissible to speculate about points over which Scripture is silent, Augustine cautions, "In this we have to strive in our own small way, to the extent that he assists us, to ensure that no absurdity or contradiction is attributed to the holy scripture."[92] If we do not work hard to harmonize knowledge acquired from general revelation with the Bible, we may incline readers of Scripture to conclude that its assertions are false, to the disastrous effect that they "decide either

[89] *De Genesi ad litteram* 6.9.14 (CSEL 28:1, 180).
[90] *De Genesi ad litteram* 9.12.22 (CSEL 28:1, 283).
[91] *De Genesi ad litteram* 2.9.21 (CSEL 28:1, 46).
[92] *De Genesi ad litteram* 5.8.23 CSEL 28:1, 152).

to give up the faith, or to have nothing to do with it in the first place."[93] Once again, Augustine's apologetic burden is evident here.

Some contemporary approaches to the doctrine of creation may benefit from considering Augustine's willingness to stand on the authority of Scripture over and against a particular scientific claim. In some accounts of recent genetic discoveries, for example, one gets the feeling that the science is setting the discussion and the theology is simply adjusting.[94] Now, as we will explore in chapter 5, I believe that the genetic data should be carefully considered, not simply shrugged off. At the same time, Augustine would remind us that it is also within the province of theology to *talk back* to science. This talking back should not, of course, be a reflexive, unstudied dismissal. But the fact that theological authority can be hastily expressed does not entail that it has no proper expression. Augustine actually approaches the Bible with the *expectation* that it will speak into our uncertainty. He regards the Bible as revelation, not mere confirmation. It is not here simply to corroborate information we would already have without it, but to shine like light into darkness, offering us certainty and truth that can be found nowhere else. In Augustine's thought, we get the impression that he both trembles before the Word of God (as much as any modern fundamentalist) and has a considered respect for science (as much as any modern progressive)—and that both of these intuitions can go *together*.[95]

---

[93]*De Genesi ad litteram* 5.8.23 (CSEL 28:1, 152).

[94]In Dennis R. Venema and Scot McKnight, *Adam and the Genome: Reading Scripture After Genetic Science* (Grand Rapids: Brazos, 2017), the authors indicate that the second part of the book (the part on theology) "assumes the correctness of the first part" (xii), which is the part on science, and then seeks to explain Adam and Eve in their ancient Near Eastern context. I have worries about this method, but I do appreciate the authors' effort at engaging genetic science and interacting with it theologically, which I will engage a bit more in chapter five.

[95]Consider, as a modern theologian who mirrors Augustine's instincts in some of these ways, Dietrich Bonhoeffer. Bonhoeffer can hardly be called a fundamentalist with respect to his views of science and Genesis 1–3, as we will mention in chapter 5; yet he retains a robust account of biblical authority. For instance, in a letter he confessed how his own approach to Scripture had developed: "The Bible alone is the answer to all our questions, and we need only to ask repeatedly and a little humbly, in order to receive this answer. One cannot simply *read* the Bible, like other

Third, stemming from this, Augustine's simultaneous respect for science and commitment to Scripture result in a more complicated view of the relationships between these two realms. On the one hand, Augustine's instincts differ from the contemporary view that the Bible occupies a completely different domain than that of science, so that its claims are limited to "spiritual" matters. This perspective is often called NOMA (non-overlapping magisteria), a term coined by Stephen Jay Gould, referring to the view that "the magisterium of science should be restricted to the empirical realm of facts, and the magisterium of religion should be restricted to questions of ultimate meaning and values."[96]

Augustine, by contrast, maintains that information from general revelation must be harmonized with what the Bible has affirmed, lest the credibility of Christianity be hindered. This presupposes that there is overlap between the interests of science and Scripture. For instance, in his commentaries Augustine expends great energy looking for ways that a particular biblical assertion can be maintained in spite of an apparent conflict.[97] In one passage, he goes on at length seeking to demonstrate the truthfulness of the assertion in Genesis 2:6 that the spring "was coming up from the earth and watering the whole face of the earth" in the face of skeptical critique. Much of book 3 of the literal commentary on Genesis is taken up with the effort of demonstrating how biblical assertions pertain to the four "classical elements" of the world (earth, water, air, and fire)[98] and the five senses of the body.[99] Augustine insists that all four elements are

books. One must be prepared really to enquire of it. Only thus will it reveal itself. Only if we expect from it the ultimate answer, shall we receive it" (as cited in Eric Metaxas, *Bonhoeffer: Pastor, Martyr, Prophet, Spy* [Nashville: Thomas Nelson, 2010], 136).

[96]J. B. Stump, "Non-Overlapping Magisteria," in *Dictionary of Christianity and Science: The Definitive Reference for the Intersection of Christian Faith and Contemporary Science*, ed. Paul Copan, Tremper Longman III, et al. (Grand Rapids: Zondervan, 2017), 488.

[97]*De Genesi ad litteram* 5.9.24-10.26 (CSEL 28:1, 152-54).

[98]E.g., *De Genesi ad litteram* 3.3.5 (CSEL 28:1, 65).

[99]E.g., *De Genesi ad litteram* 3.4.6 (CSEL 28:1, 66).

present in Genesis 1, asserting that "it is almost like a scientist that the Spirit of God, who was assisting the writer, says flying things were produced from the waters."[100] In addition, as we have seen, Augustine is perfectly happy to wield biblical assertions to counter scientific claims when he feels it necessary. Augustine is therefore out of alignment with NOMA, as well as with views like Denis Lamoureux's "message-incident principle," in which the "spiritual truths" of the Bible are distinguished from its "ancient science."[101] For all his caution against rash scientific assertions on the basis of Scripture, Augustine does not finally separate the claims of Scripture from those of science, as though "spiritual" truths and "scientific" truths can be sealed off from each other.

At the same time, Augustine is not really a concordist in the usual sense, believing that "the teaching of the Bible on the natural world, properly interpreted, will agree with the teaching of science (when it properly understands the data), and may in fact supplement science."[102] We could at most call Augustine a chastened or careful concordist, for several reasons. In the first place, as we have seen, Augustine recognizes the *spiritual purpose* of Scripture; he believes that Scripture was written primarily for our pursuit of godliness, not to help us grow in our knowledge of the natural order.[103] In line with this, Augustine emphasized the *limited scope* of Scripture. For instance, he emphasized various ways in which the biblical creation account is not exhaustive:

> Now not everything is written in scripture about how the ages ran their course after that first establishment of things, and how various stages

[100]*De Genesi ad litteram* 3.7.9 (CSEL 28:1, 66).

[101]Denis O. Lamoureux, *Evolution: Scripture and Science Say Yes!* (Grand Rapids: Zondervan, 2016), 88-91. C. John Collins observes that this approach unnecessarily equates literalism and historicity, whereas the Bible mainly employs ordinary and poetic language to describe historical events. See Collins, "Response from the Old-Earth View," in *Four Views on the Historical Adam*, ed. Matthew Barrett and Ardel B. Caneday, Counterpoints (Grand Rapids: Zondervan, 2013), 72-73. Although I do not share Lamoureux's view of relating Scripture and science, I appreciate his larger concern to help people not stumble in their faith over evolution.

[102]John Soden, "Concordism," in *Dictionary of Christianity and Science*, 104.

[103]E.g., *De Genesi ad litteram* 10.23.39 (CSEL 28:1, 326); see the discussion just above.

> followed one another in the management of creatures made at the beginning and finished on that sixth day, but only as much as the Spirit who was inspiring the author judged would be enough.[104]

As a result, Augustine held that there was a place for conjecture or speculation in creation theology: "So we in our ignorance have to fill by conjecture the gaps which he by no means out of ignorance left in the picture."[105] Perhaps this is one reason why Augustine was keen to listen attentively to the natural sciences of his day—if the Bible leaves "gaps" to be filled in, we must be open to information from other sources. But he also maintains that this must never be done in such a way as to contradict anything that Scripture teaches.

Above all, Augustine encourages a more tentative posture in seeking harmony between Scripture and science.[106] We have seen how he is constantly worrying about "rashness," and is content to admit what he does not know. He even displays a sensitivity toward tactful expressions that will not unnecessarily embarrass or provoke his readers. For example, after offering a possible theory about the origins of light, he backpedals: "But if I say that I am afraid I will be laughed at by those who know for certain."[107] Such language reflects a carefulness that is not always present among strong concordists.

If Augustine is not neatly in either the NOMA or concordist camp, where does he fit? A different way of approaching the relation of Scripture and science is pictured in Derek Kidner's commentary on Genesis:

> The accounts of the world [of science and Scripture] are as distinct (and each as legitimate) as an artist's portrait and an anatomist's diagram, of which no composite picture will be satisfactory, for their common ground is only in the total reality to which they both

---

[104]*De Genesi ad litteram* 5.8.23 (CSEL 28:1, 152).
[105]*De Genesi ad litteram* 5.8.23 CSEL 28:1, 152).
[106]Such carefulness is well modeled by Alvin Plantinga in *Where the Conflict Really Lies: Science, Naturalism, and Religion* (Oxford: Oxford University Press, 2011). Plantinga is particularly helpful in pushing back against unwarranted philosophical extrapolations from scientific claims.
[107]*De Genesi ad litteram* 1.10.21 (CSEL 28:1, 15).

> attend. . . . [Scripture's] bold selectiveness, like that of a great painting, is its power.[108]

This view assigns greater complexity to the relation of science and Scripture, and more distinctness between them, than concordist approaches usually do. (Some kind of "composite picture" is generally the very goal of concordism.) At the same time, it is not exactly NOMA, since a diagram and a portrait can overlap in speaking of the same material content. For instance, it is not at all obvious that a painting conveys only meaning/values and no facts, or that a diagram conveys only facts and no meaning/values. Kidner's metaphor is not even identical with the intermediate view, so-called TOMA (tangentially overlapping magisteria),[109] since a painting and a diagram are radically different methods of representing truth, while TOMA still has in view two methodologically similar approaches to truth (hence the word "magisteria").

The painting/diagram image does not fully encapsulate Augustine's approach to science-faith dialogue. But his carefulness and flexibility about the relation of science and Scripture, his openness to potential conflict between the two, and his preservation of the Bible's ability to speak with power all encourage the more complicated, chastened approach that is entailed by Kidner's metaphor.

DOES HUMILITY MEAN A LACK OF CONVICTION?

In closing, we may consider a possible objection that some readers might have wondered about at points throughout this chapter. Someone could argue, "It is all too easy to advocate for humility in areas you consider less weighty. But if the issue at hand is the deity of Christ, you wouldn't be warning about the dangers of rashness!"

---

[108]Derek Kidner, *Genesis: An Introduction and Commentary*, The Bible Speaks Today (Downers Grove, IL: InterVarsity Press, 1967), 31.

[109]This term is drawn from William P. Brown, *The Seven Pillars of Creation: The Bible, Science, and the Ecology of Wonder* (Oxford: Oxford University Press, 2010); see the insightful discussion in J. Richard Middleton, "Reading Genesis 3 Attentive to Human Evolution," in *Evolution and the Fall*, 70-72.

This is a valid concern. In response, we must recognize a point that has been somewhat implicit throughout what we have already surveyed: for Augustine, humility was not the opposite of conviction, as though in order to be humble one must adopt a vaguely deferential mindset on every issue. As we have seen, Augustine made a principled distinction between the clear/central aspects of creation, on the one hand, and the relatively murky/peripheral, on the other—what he called "certainties" versus "opinions." For him, humility entailed an unflinching allegiance to the former as much as a prudential discretion about the latter.[110]

To give one example of this, at the start of his unfinished commentary on Genesis Augustine advocates for a questioning posture toward the doctrine of creation, particularly when it comes to the interpretation of Scripture:

> The obscure mysteries of the natural order, which we perceive to have been made by God the almighty craftsman, should rather be discussed by asking questions than be making affirmations. This is supremely the case with the books that have been entrusted to us by divine authority, because the rash assertion of one's uncertain and dubious opinions in dealing with them can scarcely avoid the charge of sacrilege.[111]

All this is quite familiar by now. But then, after this statement, Augustine proceeds to warn that there are boundaries within which this questioning posture should not stray: "On the other hand the doubts and hesitations implied by asking questions must not exceed the bounds of Catholic faith."[112] Augustine is aware that many heretics have twisted these Scriptures to fit their own ideas, and their views

---

[110]This way of thinking about humility as a theological virtue accords with the insight of G. K. Chesterton, *Orthodoxy* (Garden City, NY: Doubleday, 1957): "What we suffer from today is humility in the wrong place. Modesty has moved from the organ of ambition. Modesty has settled upon the organ of conviction; where it was never meant to be. A man was meant to be doubtful about himself, but undoubting about the truth; this has been exactly reversed. Nowadays the part of a man that a man does assert is exactly the part he ought not to assert—himself" (31).

[111]*De Genesi ad litteram liber unus imperfectus* 1.1 (CSEL 28:1, 459).

[112]*De Genesi ad litteram liber unus imperfectus* 1.1 (CSEL 28:1, 459).

must be opposed. Therefore, he opens his commentary with a summary of the Catholic faith, in the form of the Apostles' Creed in its African variation (with occasional references to the Nicene Creed).[113] The particular areas he emphasizes as inviolable are trinitarian agency in creation, creation's non-eternality, the goodness of creation, and the redemption of creation through the work of Christ.[114] This is similar to his approach in *The City of God*, where he develops his doctrine of the knowledge of God through Christ the Mediator (11.2) and his doctrine of Scripture (11.3) up front before launching into his treatment of the origin of the two cities.[115]

If more treatments of creation made the Apostles' Creed our emotional center of gravity, perhaps our dialogue with other Christian views would be more irenic and our engagement with non-Christian ideas simultaneously more incisive.

For Augustine, then, humility within the doctrine of creation involves this kind of methodologically self-conscious balance, in which we are as eager to affirm the weighty matters of orthodoxy as we are circumspect in our private judgments about the more contested areas. If it is a violation of humility to be dogmatic about the fringes, it is equally so to waver at the center.

In other words, to put it as briefly and punchily as possible: humility does not mean saying "I don't know" to every question. It means saying "I don't know" when, in fact, you don't.

---

[113] *De Genesi ad litteram liber unus imperfectus* 1.1 (CSEL 28:1, 459). The African form of the Apostles' Creed was the baptismal creed of the African church. See the discussion in Augustine, *On Genesis*, 116.

[114] *De Genesi ad litteram liber unus imperfectus* 1.2-4 (CSEL 28:1, 459-61).

[115] *De civitate Dei* 11.2-3 (CSEL 40.1, 512-14).

CHAPTER THREE

SETTLING AN AGE-OLD DEBATE

Augustine on the Literal Meaning of Genesis 1

One must take care not to interpret a figurative expression literally. What the apostle says is relevant here: "the letter kills, but the spirit gives life" [2 Cor 3:6]. . . . It is, then, a miserable kind of spiritual slavery to interpret signs as things, and to be incapable of raising the mind's eye above the physical creation so as to absorb the eternal light.

De Doctrina Christiana 3.5.9

Different views on creation in the church today are often summarized in terms of whether one takes the biblical creation story "literally." Augustine's crowning achievement on the doctrine of creation was likewise, as we have noted, the production of a "literal" commentary on Genesis 1–3.

Yet what Augustine means by "literal" is quite different from many modern uses of this term. When modern Christians (and, in many cases, non-Christians) speak of a "literal" interpretation of Genesis 1, they often have in mind the claim that the six days depicting God's work of creation in Genesis 1 represent twenty-four-hour, sequential, non-overlapping periods of time, such that the world was created in a space of roughly 144 hours, generally around 6,000–10,000 years ago. This is what it means, we are told, to read the text "literally."

Several other claims are often bundled together with this one. First, this "literal" interpretation is set forth as the "historical" reading of the text. For instance, a recent film promoting this interpretation of Genesis 1 presents two options: full-blown naturalism or the "historical Genesis paradigm" (young-earth creationism). The film impresses on its viewers a choice: either the days are twenty-four-hour periods of time, or we have denied the text's historicity. Indeed, since alternative views of creation are not even mentioned, many viewers will likely conclude that unless the universe was created recently, there is no possibility of divine involvement in the work of creation at all.[1]

Second, this "literal" interpretation of Genesis 1 is often presented as the only view that honors the authority of Scripture and protects the integrity of the gospel. For example, one piece of promotional literature from a young-earth creationist ministry offers the warning, "what is at stake is nothing less than the authority of Scripture, the character of God, the doctrine of death, and the very foundation of the gospel. If the early chapters of Genesis are not true literal history, then faith in the rest of the Bible is undermined, including its teaching about salvation and morality."[2] Once again, "true literal history" in this context means twenty-four-hour days and a recent creation.

Finally, this reading of Genesis 1 is often portrayed as the unbroken consensus of premodern Christianity, such that any alternative view is essentially motivated by compromise with the claims of modern science. For example, some advocates for this view claim that "no significant debate existed on the matter before the nineteenth century because the plainest and most straightforward reading of the text had

---

[1]*Is Genesis History?*, directed, written, and produced by Thomas Purifoy and narrated by Del Tackett, initially released in limited theaters in early 2017.

[2]Ken Ham, *A Pocket Guide to Compromise: Refuting Non-Biblical Interpretations of Genesis 1* (Petersburg, KY: Answers in Genesis, 2011), 7.

no sustained challengers."[3] Adherence to this "literal" interpretation is therefore portrayed as simply maintaining what all faithful Christians have always believed: "We side with the Law and the Prophets. We side with the apostles. We side with the consensus of the Church fathers. We side with the Reformation and Puritan divines. We side with the uniform testimony of the Church until recently. We can do no other."[4]

Up until now, we have sought Augustine's help for both widening our vision of creation and cautioning our method. In this chapter, we will enlist him to weigh in more directly on a current area of dispute: What does it mean to read Genesis 1 *literally*?[5] Augustine is helpful on this question because he devoted enormous energy to pursuing a "literal" meaning of the text, producing multiple commentaries over decades of reflection and struggle. Moreover, as we shall see, the view he arrived at had a significant influence on subsequent interpreters, especially throughout the medieval era. The legacy of Augustine's exegesis of Genesis 1–3 punctures the young-earth creationists' claim that their view is tantamount to that of premodern Christianity. Nonetheless, here we are more interested in the *reasons* Augustine arrived at his view and how they may invite us to consider the complexities involved in the interpretation of this passage—as well as the need for charity, humility, and precision in how we discuss its meaning in the church today.

---

[3]J. Ligon Duncan III and David W. Hall, "The 24-Hour View," in David G. Hagopian, ed., *The Genesis Debate: Three Views on the Days of Creation* (Mission Viejo, CA: Cruxpress, 2001), 22. Later they write: "This straightforward understanding is . . . the consensus of the greatest minds in the history of Christianity," and alternative views find "no support in the witness of history" (29). They also claim that "to embrace the framework interpretation, you have to believe that everybody in the history of interpretation got it wrong—badly wrong—until [Meredith] Kline" (267).

[4]Duncan and Hall, "The 24-Hour View," 60.

[5]Augustine's approach to Scripture as a whole is too complicated a topic to cover fully here. For a fuller treatment of his approach to Scripture and how it interfaces with modern approaches, see Michael Graves, *The Inspiration and Interpretation of Scripture: What the Early Church Can Teach Us* (Grand Rapids: Eerdmans, 2014).

AUGUSTINE'S JOURNEY FROM ALLEGORICAL TO LITERAL

Augustine did not start with a literal interpretation of Genesis. As we have seen, his first effort at a commentary on the creation account was a two-volume allegorical interpretation written specifically against the Manichaeans. This included such ideas as taking the days of Genesis 1 as seven epochs of redemptive history and seven stages of the Christian life.[6] This redemptive-historical interpretation of Genesis 1 also characterizes Augustine's sermons, where he argues that each day represents a long "age" of human history:

> There will be a Sabbath of this world, when six ages have passed. The age are passing like six days. One day, from Adam to Noah, has passed; another from the flood to Abraham, has passed; a third, from Abraham to David, has passed; a fourth, from David to the exile in Babylon, has passed; a fifth, from the exile in Babylon to the coming of our Lord Jesus Christ. Now the sixth day is being spent. We are in the sixth age, the sixth day.[7]

When Augustine described his later works on Genesis as "literal," he employed this term to distinguish them from his earlier allegorical approach. Thus, in his *Retractations*, he defines the word *literal* in the title *The Literal Commentary on Genesis* as meaning "not the allegorical meanings of the text, but the proper assessment of what actually happened."[8] At the same time, in attempting a literal commentary Augustine was not rejecting allegorical interpretation wholesale. Instead, as Yoon Kyung Kim has argued at length, in the course of his development Augustine's understanding of the literal meaning progressed to encompass the allegorical as well.[9]

---

[6] *De Genesi contra Manichaeos* 1.23.35-25.43 (CSEL 91, 104-14).

[7] Sermon 125.4, in *Sermons 94A–147A on the New Testament*, ed. John E. Rotelle, trans. Edmund Hill, The Works of Saint Augustine: A Translation for the 21st Century (Brooklyn, NY: New City Press, 1992), 256.

[8] *Retractationes* 2.50 (CSEL 36, 159).

[9] Yoon Kyung Kim, *Augustine's Changing Interpretations of Genesis 1–3: from De Genesi contra Manichaeos to De Genesi ad litteram* (Lewiston, NY: The Edwin Mellen Press, 2006), 163-67.

This explains why we find Augustine continuing to advocate for various allegorical interpretations in the literal commentaries. In the very first paragraph of his finished literal commentary, for instance, he affirms multiple levels of meaning in Scripture, and asks whether texts recounting history should be taken "as only having a figurative meaning, or whether they are also to be asserted and defended as a faithful account of what actually happened."[10] In developing his answer to this question, Augustine acknowledges that both figurative and historical meanings can be present in a text—but strikingly, it is the figurative meaning that is assumed and the historical that needs to be established. He grounds this assumption with an appeal to the apostle Paul. After claiming that "no Christian . . . will have the nerve to say that [the early chapters of Genesis] should not be taken in a figurative sense," he quotes Paul's endorsement of figurative language in 1 Corinthians 10:11 and Ephesians 5:32.[11] Then he proceeds to affirm that "the text has to be treated in both ways," and offers literal interpretations of the succeeding verses in addition to allegorical ones.[12] Similarly, early on in his unfinished literal commentary he affirms the validity of allegorical interpretation alongside historical, analogical, and etiological interpretation (following a fourfold hermeneutic developed by others).[13] All this suggests that Augustine does not regard the literal meaning as superseding or replacing the allegorical.

Furthermore, throughout these literal commentaries Augustine remains comfortable with various allegorical interpretations. In fact, he often reaffirms specific allegorical interpretations contained in his earlier works. For example, on the question of whether the sky is more like a dome or a lid, he complains about "the tiresome people who persist in demanding a literal explanation" and directs his

---

[10] *De Genesi ad litteram* 1.1.1 (CSEL 28:1, 3).
[11] *De Genesi ad litteram* 1.1.1 (CSEL 28:1, 3).
[12] *De Genesi ad litteram* 1.1.2 (CSEL 28:1, 3).
[13] *De Genesi ad litteram liber unus imperfectus* 2.5 (CSEL 28:1, 461).

readers to an allegorical interpretation he had offered earlier: "My treatment of this in terms of allegory may be found in the thirteenth book of my *Confessions*."[14]

Augustine's translators and commentators have not failed to notice the irony of his protests against *literalism* in the midst of a *literal* commentary. Edmund Hill, for instance, notes that "this remark here is very revealing about his real intentions, and his thoroughly 'anti-fundamentalist' understanding of the literal sense. Both these comparisons of the sky, or heavens, to a skin stretched out and to a dome or vault are clearly poetic, metaphorical, not intended by the author to be taken as literal descriptions in the narrow sense."[15]

In addition, we find Augustine continuing to advance allegorical readings of Genesis 1 in his sermons well into the 400s.[16] And beyond all this, allegorical interpretation of Genesis 1 recurs in Augustine's other works, such as *Confessions*, which Augustine wrote after his first attempt at a literal commentary. Thus, Augustine's movement toward a literal reading was a development within an overarching continuity, and the approach to the text that he ultimately landed on involved more than simply replacing his earlier methods. Ultimately, Augustine's conception of the text's literal meaning involved various levels of meaning, including an allegorical sense.[17]

Part of the rationale for pitting Augustine's allegorical and literal interpretations against one another comes from a particular understanding of Augustine's development as a theologian that Carol Harrison refers to as the "revolution of the 390s." This idea, popularized in Peter Brown's influential 1967 biography of Augustine and somewhat standard in Augustine scholarship since that time, posits a fairly sharp

---

[14]*De Genesi ad litteram* 2.9.22. (CSEL 28:1, 47).

[15]Augustine, *On Genesis*, ed. John E. Rotelle, trans. Edmund Hill, The Works of Saint Augustine: A Translation for the 21st Century (Hyde Park, NY: New City Press, 2002), 202.

[16]E.g., Sermon 229V, in *Sermons 184–229Z on the Liturgical Seasons*, ed. John Rotelle, trans. Edmund Hill, The Works of Saint Augustine: A Translation for the 21st Century (Hyde Park, NY: New City Press, 1993), 340. See Rotelle's discussion of the sermon's date and authenticity on 341.

[17]This point is established more fully by Kim, *Augustine's Changing Interpretations of Genesis 1–3*.

discontinuity between the more optimistic, philosophically inclined Augustine following his conversion in 386 and the more pessimistic, dogmatically oriented Augustine following his study of Paul in the early 390s.[18] As a result, according to this thesis, during the decade between his conversion in 386 and his consecration as bishop in 396 (around the time he began to write the *Confessions*), the Neoplatonic influence somewhat receded and Augustine hardened into his mature views on the fall, original sin, divine grace, and predestination.

Harrison, however, has marshaled evidence that the central pillars of Augustine's mature theology were in place from 386 onward, and that the separation of his early thought as a Christian from his "mature" theology ultimately amounts to a caricature.[19] In an epilogue to the new edition to his biography entitled "New Directions," Brown himself notes that he strove to portray "a sense of human movement in a figure usually identified with all that was most rigid and unmoving in Catholic dogma."[20] We must appreciate Brown's even-handed acknowledgment that

> such an emphasis on the changes in Augustine's thought and outlook can be challenged. Central elements in Augustine's thought have been shown to be remarkably stable. They seem to bear little trace of discontinuity. Augustine's intellectual life as a bishop cannot be said to have been lived out entirely in the shadow of a "Lost Future," as I had suggested in the chapter of my book which bears that title. In the same manner, the later decades of Augustine's thought on grace, free will and predestination cannot be lightly dismissed as the departure of a tired old man from the views of an earlier, "better" self. As a thinker, Augustine was, perhaps, more a man *aus einem Guss*, all of a piece, and less riven by fateful discontinuities than I had thought.[21]

[18]See especially the chapter "The Lost Future" in Peter Brown, *Augustine of Hippo: A Biography*, rev. ed. (Los Angeles: University of California Press, 2000).

[19]Carol Harrison, *Rethinking Augustine's Early Theology: An Argument for Continuity* (Oxford: Oxford University Press, 2006), 1-7, 14-19. As she puts it, succinctly and ironically: "Augustine's early thought was not only fully Christian; it was also fully Augustinian" (286).

[20]Brown, *Augustine of Hippo*, 490.

[21]Brown, *Augustine of Hippo*, 490.

We cannot here resolve all of the challenging issues involved in questions of continuity/discontinuity in Augustine's postconversion development. But for our purposes we would suggest at the very least that there is no basis for an a priori exclusion of Augustine's earlier writings (such as *De Genesi contra Manichaeos*) from consideration of a mature account of his doctrine of creation, and that, more generally, Augustine's literal interpretation of the text should not be pitted against his allegorical interpretation.

But if *literal*, as Augustine intends the term, means something different than simply "not allegorical," what *does* it mean? To put the matter simply: for Augustine, the term *literal* was concerned with historical referentiality, not with the particular manner in which that history was recounted. In other words, Augustine did not employ the word *literal* to designate one literary style or genre of historical narration over another, but to designate historical narration as such. Thus, he defines *literal* in the literal commentary as "a faithful account of what actually happened,"[22] and in the *Retractations* as "the proper assessment of what actually happened."[23] As different as this is from how we sometimes use the term today, Augustine was not unusual among the church fathers for thinking about literality in this way.[24]

In fact, even today the word *literal* involves a complex range of meaning, some instances of which mirror Augustine's usage. For instance, Kevin Vanhoozer describes a good/soft literality, distinct from a hard/bad literality, as an interpretation that is "sensitive to the way language works, and acknowledges intended figures of speech as part and parcel of the literal sense."[25] Thus, describing a statement or

---

[22]*De Genesi ad litteram* 1.1.1 (CSEL 28:1, 3).

[23]*Retractationes* 2.50 (CSEL 36, 159).

[24]E.g., see Graves, *Inspiration and Interpretation of Scripture*, 62-63.

[25]Kevin J. Vanhoozer, "Shining Light on Literality: From the Literal Interpretation of Genesis to the Doctrine of Literal Six-Day Creation," lecture given at the Carl F. H. Henry Center for Theological Understanding, September 14, 2016, http://henrycenter.tiu.edu/resource/from-the-literal-interpretation-of-genesis-to-the-doctrine-of-literal-six-day-creation. For Vanhoozer's fuller and masterful account of biblical hermeneutics, see *Is There a Meaning in This Text? The*

piece of literature as "literal" need not exclude the possibility of language that is metaphorical, figurative, pictorial, dramatic, stylized, or poetical. For example, if someone says, "It's raining cats and dogs outside," there is a sense in which supposing animals are falling from the sky is precisely a *failure* to take this statement "literally."

Augustine's literal commentaries display a sensitivity to these kinds of literary considerations. He will often pause to remind his readers of a linguistic consideration or offer them a warning to heed how the Bible is using a particular figure of speech (*figura locutionis*).[26] It is not uncommon to find him pausing to worry whether an interpretation he has just advanced is not, in fact, an "altogether absurd and literal-minded, fleshly train of thought,"[27] or to receive a warning from him that we "never think in a literal-minded, fleshly way of utterances" in regard to the creation days.[28] For Augustine, taking Scripture in an overly literal sense is not just a hermeneutical error but a spiritual one—he describes this danger as "fleshly" and a mark of spiritual immaturity, deserving of Paul's admonition in 1 Corinthians 14:20, "Do not be children in your thinking."[29] Accordingly, he will severely berate the Manichaeans for their "crude, literal-minded" way of thinking about God, and the "sacrilegious mumbo-jumbo" involved in their descriptions of his activity.[30] Elsewhere, when dealing with the assertion in Sirach that "the sun rises, and the sun sets, and leads to its place," Augustine rejects as a "monstrous supposition" what he terms the "poetic fiction" of the classic myth (probably represented

Bible, the Reader, and the Morality of Literary Knowledge, Landmarks in Christian Scholarship (Grand Rapids: Zondervan, 1998).

[26]E.g., *De Genesi contra Manichaeos* 1.22.34 (CSEL 91, 102-4); for another example, cf. his discussion of the figure of speech employed in the phrase "he has become like one of us" in Genesis 3:22 (*De Genesi contra Manichaeos* 2.22.33 [CSEL 91, 101-2]).

[27]*De Genesi ad litteram* 1.2.5 (CSEL 28:1, 6).

[28]*De Genesi ad litteram* 1.18.36 (CSEL 28:1, 27).

[29]*De Genesi ad litteram* 1.18.36 (CSEL 28:1, 27); cf. his similar warning in *De Genesi contra Manichaeos* 1.22.33 (CSEL 91, 101-2).

[30]*De Genesi ad litteram* 7.11.17 (CSEL 28:1, 210-11). "Sacrilegious mumbo-jumbo" could be more plainly rendered as "vain and sacrilegious judgments."

to him in its depiction in Ovid's *Metamorphoses*) that "the sun sinks into the sea and rises, well washed, on the other side."[31]

Being overly literal is not, of course, the only error that Augustine worries about. He is also wary of too quickly leaving off of a literal interpretation. That is why he attempts the literal commentary in the first place. In *both* his allegorical and his literal works on Genesis, in fact, Augustine stipulates that we should look for literal interpretations of the narrative of Genesis whenever we can, and fall back on an exclusively figurative meaning only when necessary. Thus it is in his allegorical commentary that he gives the following warning:

> One should not look with a jaundiced eye, to be sure, on anyone who wants to take everything that is said here literally, and who can avoid blasphemy in so doing, and present everything as in accordance with Catholic faith. . . . If, however, no other way is available of reaching an understanding of what is written that is religious and worthy of God, except by supposing that it has all been set before us in a figurative sense and in riddles, we have the authority of the apostles for doing this, seeing that they resolved so many riddles in the books of the Old Testament in this manner.[32]

That Augustine can make this assertion in his allegorical commentary, just as he will later affirm allegorical interpretation in his literal work, demonstrates once again that he does not regard these two hermeneutical approaches as mutually exclusive.

Augustine's perception of literalism as a *spiritual* matter is related to his testimony as recounted in the *Confessions*. After describing the impression that Ambrose's rhetorical skills as a preacher had on him, surpassing as they did those of the Manichaean leader Faustus, Augustine narrates how he gradually softened toward the content of Ambrose's sermons. In this connection, he emphasizes particularly

---

[31]*De Genesi ad litteram* 1.10.21 (CSEL 28:1, 16). He calls this a "poetic fiction." On his debt to Ovid, see Hill's comments in *On Genesis*, 177.

[32]*De Genesi contra Manichaeos* 2.2.3 (CSEL 91, 120-21).

Ambrose's ability to explain Old Testament passages such as Genesis 1 in a figurative way:

> I began to see that the Catholic faith, for which I had thought nothing could be said in the face of the Manichaean objections, could be maintained on reasonable grounds: this especially after I had heard explained figuratively several passages of the Old Testament which had been a cause of death for me when taken literally. Many passages of these books were expounded in a spiritual sense and I came to blame my own hopeless folly in believing that the law and the prophets could not stand against those who hated and mocked at them.[33]

Later he expresses his relief in discovering, with Ambrose's help, that his objections were not against the Scripture itself, but against the ideas his own over-literalism had mistakenly forced onto the Scripture.[34] The concerns reflected in Augustine's own experience, evident in this passage, reverberate throughout his exegesis of Genesis 1, even in his efforts at a literal commentary.

AUGUSTINE'S VIEW OF THE CREATION DAYS

So what does Augustine think Genesis 1 *literally* means? This is a different question than asking how old Augustine thought the universe was. It is well known that Augustine claimed that humanity is less than six thousand years old,[35] and young-earth creationists have made much of this fact.[36] But it is problematic to use Augustine's affirmation of a recent humanity as a support for a "literal" (in the young-earth creationist sense) reading of Genesis 1, for several reasons.

First, Augustine did not derive his view of creation's age from a "literal" reading of the days in Genesis 1 in the way young-earth creationists do,

[33] *Confessiones* 5.14 (CSEL 33, 111); translation from F. J. Sheed, *Confessions*, 90-91.

[34] *Confessiones* 6.3-4 (CSEL 33, 116-20).

[35] *De civitate Dei* 12.12 (CSEL 40.1, 585).

[36] E.g., James R. Mook, "The Church Fathers on Genesis, the Flood, and the Age of the Earth," in *Coming to Grips with Genesis: Biblical Authority and the Age of the Earth*, ed. Terry Mortenson and Thane H. Ury (Green Forest, AR: Master Books, 2008), 35-39.

so the association is somewhat artificial (in his day, Augustine had no particular reason to envision an older universe). Second, Augustine is speaking of the age of humanity, not the age of the universe. Augustine, in fact, nowhere stipulates how old the world itself is. James Mook counters this claim with an appeal to Augustine's doctrine of instantaneous creation: "Lest it be argued on the basis of Augustine's statement that Adam was created less than 6,000 years ago but the rest of creation is much older than that, it must be remembered Augustine believed that God created everything, at least seminally, in an instant."[37] But Augustine did not believe the creation of Adam was a part of the instantaneous creation—he believed it came later,[38] and he nowhere offers a time frame for how much time elapsed between the instantaneous creation moment and the creation of humanity. In fact, he expressly claims ignorance on this point just a few pages after the passage in which he affirms a recent humanity: "I own that I do not know what ages passed before the human race was created."[39] Augustine is not necessarily thinking here of long earthly epochs; we must not read old-earth creationist visions of long stretches of time back onto him. But we may also not claim that he dates the universe; in fact, his treatment of the question, which is bound up with whether God was ever not Lord of something, emphasizes the difficulty of this issue, and the danger of raising "hazardous questions."[40]

A third problem with associating Augustine's view of the age of humanity with the young-earth creationist kind of literalism is that in the very context of his assertion that humanity is recent, the whole burden of his argument turns on the principle that our intuitions about the timing of creation are not reliable:

---

[37]Mook, "Church Fathers on Genesis," 38.

[38]*De Genesi ad litteram* 6.2.3-3.5 (CSEL 28:1, 171-73); we will discuss this point further in chapter 5.

[39]*De civitate Dei* 12.16 (CSEL 40.1, 595).

[40]*De civitate Dei* 12.15 (CSEL 40.1, 594).

> If there had elapsed since the creation of man, I do not say five or six, but even sixty or six hundred thousand years, or sixty times as many, or six hundred or six hundred thousand times as many, or this sum multiplied until it could no longer be expressed in numbers, the same question could still be put, Why was he not made before? For the past and boundless eternity during which God abstained from creating man is so great, that, compare it with what vast and untold number of ages you please, so long as there is a definite conclusion of this term of time, it is not even as if you compared the minutest drop of water with the ocean that everywhere flows around the globe.[41]

Augustine has the Manichaean objection in view here, but his response is based on the more basic claim that the comparison between eternity and time far outweighs any relative time lengths. One can well imagine how an old-earth appeal can be made from the *principle* of Augustine's view of age as much (or better) than a young-earth appeal from its material claim, for Augustine seems to think that the sheer passage of time prior to humanity is irrelevant to the more basic philosophical conundrums involved with the notion of creation itself.

Now to return to the first point: What does Augustine believe about the days of creation? In his finished literal commentary on Genesis, he emphasizes the ineffability of the creation act and the difficulty in accessing its meaning: "It is indeed an arduous and extremely difficult task for us to get through to what the writer meant with these six days, however concentrated our attention and lively our minds."[42] He cautions against hasty theorizing about the nature of the days of Genesis 1 since we have no direct access to the creation act:

> Now clearly, in this earth-bound condition of ours we mortals can have no experiential perception of that day, or those days which were named numbered by the repetition of it; and even if we are able to struggle towards some understanding of them, we certainly ought not

[41]*De civitate Dei* 12.12 (CSEL 40.1, 585).
[42]*De Genesi ad litteram* 4.1.1 (CSEL 28:1, 93).

> to rush into the assertion of any ill-considered theory about them, as if none more apt or likely could be mooted.[43]

Augustine's worry about rashness, surveyed in the last chapter, is evident here. Once again, this very caution might rub against the claim that interpreting the days of Genesis 1 is a matter of obviousness. If we insist that the right approach to the text is like a wide road in broad daylight, only ever willfully refused and never simply misunderstood, then it is difficult not to attribute either stupidity or deviousness to the greatest theologian of the early church.

Ultimately, in developing his own view of the creation days, Augustine emphasizes the difference of the days of Genesis 1 from ordinary days, and the uniqueness of the creation event. He is willing to grant, for the sake of argument, that our twenty-four-hour days "represent those first seven in some fashion," but he hastens to add that "we must be in no doubt that they are not at all like them, but very, very dissimilar."[44] Later, Augustine will assert that ordinary twenty-four-hour days are a kind of "shadow" (*umbram*) of the original creation days.[45] This language might suggest that Augustine holds something like a "day age" view, in which sequence remains an important feature of the account. But as it turns out, Augustine argues repeatedly that the creation events depicted in Genesis 1 occurred in one instantaneous moment. Thus, Augustine not only distinguishes the days of Genesis 1 from ordinary twenty-four-hour days, he also distinguishes God's initial creative act (which he regards to be instantaneous) from his subsequent activity in creation:

> When we reflect upon the first establishment of creatures in the works of God from which he rested on the seventh day, we should not think either of those days as being like these ones governed by the sun, nor of that working as resembling the way God now works in time; but we

---

[43]*De Genesi ad litteram* 4.27.44 (CSEL 28:1, 126).
[44]*De Genesi ad litteram* 4.27.44 (CSEL 28:1, 126).
[45]*De Genesi ad litteram* 4.33.56 (CSEL 28:1, 136).

should reflect rather upon the work from which times began, the work of making all things at once, simultaneously.[46]

But if God created all things simultaneously, why is this act described in terms of the six days of a human work week? Augustine maintains that the six-day construct in Genesis 1 is an accommodation in which "the scriptural style comes down to the level of the little ones and adjusts itself to their capacity."[47] Augustine has a thoroughgoing appreciation of the notion of accommodation—the idea that God has adjusted his revelation so as to be comprehensible to the particular people to whom he is communicating.[48] It is an idea that recurs throughout his commentaries. Elsewhere, for instance, he will describe Scripture as speaking "in a weak and simple style" when communicating to the weak and simple,[49] or expressing invisible realities with visible names "out of consideration for the weakness of the little ones."[50] In yet another place, he compares biblical language to a mother matching her steps to those of a toddler learning to walk.[51]

Augustine believes that accommodation is needed to depict creation because the act of creation is utterly beyond human understanding. Responding to the overly literal Manichaean misunderstanding of Genesis 1:4, he maintains, "But so that we should be suitably brought up and helped to attain to those things that cannot be uttered by any human speech, things are said in scripture which we are able to grasp."[52] Elsewhere, he wonders whether perhaps the

---

[46]*De Genesi ad litteram* 5.5.12 (CSEL 28:1, 145-46).

[47]*De Genesi ad litteram* 2.6.13 (CSEL 28:1, 41).

[48]Augustine's conception of accommodation, unlike later variations of accommodation associated with the rise of higher criticism, does not involve the notion that God allows errors into the Bible's assertions and worldview. For an overview of how the notion of accommodation played out in the modern era, particularly in Dutch and German contexts, see Hoon J. Lee, *The Biblical Accommodation Debate in Germany: Interpretation and the Enlightenment* (Basingstoke, UK: Palgrave Macmillan, 2017). I am grateful to John Woodbridge for directing me to Lee's work.

[49]*De Genesi ad litteram* 5.6.19 (CSEL 28:1, 148).

[50]*De Genesi contra Manichaeos* 1.5.9 (CSEL 91, 76).

[51]*De Genesi ad litteram* 5.3.6 (CSEL 28:1, 141).

[52]*De Genesi contra Manichaeos* 1.8.14 (CSEL 91, 80).

days of Genesis 1 were arranged in the way they were "as a help to human frailty, and to suggest sublime things to lowly people in a lowly manner by following the basic rule of story-telling, which requires the story teller's tale to have a beginning, a middle, and end."[53] That Augustine can depict Genesis 1 as a "lowly manner" of storytelling is striking in itself—that he can categorize this description as a "literal" interpretation even more so. In fact, he is emphatic that his interpretation of Genesis 1 does not constitute a figurative, allegorical, or metaphorical reading of the text.[54]

So what kind of accommodation, specifically, does Augustine see in the days of Genesis 1? Ultimately, he affirms that the ordering of Genesis 1 is not according to temporal sequence but rather the ordering of angelic knowledge.[55] Angels play a profound role in Augustine's theology generally. He uses the phrase "heaven of heavens" to refer to the abode of angels; it is essentially an intellectual place; it is the "wisdom" of Proverbs; it is a place of light.[56] Augustine spends a lot of energy in his literal commentary developing his views on angels, and specifically on angelic foreknowledge. At one point he even speculates, drawing from texts like Ephesians 3:8-11 and 1 Timothy 3:16, "It would be surprising, unless I am very much mistaken, if God is ever said to know anything at a particular moment in time, without its really meaning that he causes it to be known, whether by angels or by human beings."[57]

With respect to their role in creation, Augustine suggests that angels are the light of Genesis 1:3. Thus he writes, "On the first day, on which light was made, the setting up of the spiritual and intelligent creation is being announced under the name of light—the nature of

---

[53]*De Genesi ad litteram liber unus imperfectus* 3.8 (CSEL 28:1, 463).

[54]*De Genesi ad litteram* 4.28.45 (CSEL 28:1, 126-27).

[55]E.g., *De Genesi ad litteram* 4.25.56, (CSEL 28:1, 124).

[56]For a helpful discussion, see Rowan Williams, "Creation," in *Augustine Through the Ages: An Encyclopedia*, ed. Allan D. Fitzgerald (Grand Rapids: Eerdmans, 1999), 253.

[57]*De Genesi ad litteram* 5.19.39 (CSEL 28:1, 163).

this creation being understood to include all the angels and powers."[58] Augustine has previously asserted that all created things exist in the Son as the Wisdom of the Father before they are actually made.[59] But now he develops a three-tiered account of the logical order of creation: first, created things are in the Word; second, they are in the angelic knowledge; and finally, they come into actual existence. Thus, with reference to the creation of heaven (not the spiritual realm, the "heaven of heavens," but rather the solid structure called heaven on day two of creation), Augustine writes:

> The fashioning of heaven, on the other hand, or the sky, was first made in the Word of God in terms of begotten Wisdom, then it was made next in the spiritual creation, that is, in the knowledge of the angels, in terms of the wisdom created in them, and only next after that was the heaven made, so that the actual created heaven might be there in its own specific kind.[60]

Augustine proceeds to stipulate that this same threefold order (in the Word, then in angelic knowledge, then in actual existence) is true of all the other elements created in Genesis 1. This must be the case, Augustine thinks, because angels "are enlightened in order to live wisely" by the Word of God.[61] From the moment they exist, they ceaselessly enjoy the Word in "holy and devout contemplation," and see material objects in him.[62] Thus, if created objects are in the Word, they must be known by those angels who peer ceaselessly into the Word. As Augustine puts it, "Just as the formula or idea on which a creature is fashioned is there in the Word of God before it is realized in the fashioning of the creature, so also is knowledge of the same formula or idea first produced in the intelligent creation," (i.e., the angels).[63] Thus,

---

[58]*De Genesi ad litteram* 2.8.16 (CSEL 28:1, 43).
[59]E.g., *De Genesi ad litteram* 2.6.12 (CSEL 28:1, 41).
[60]*De Genesi ad litteram* 2.8.16 (CSEL 28:1, 43-44).
[61]*De Genesi ad litteram* 2.8.16 (CSEL 28:1, 44).
[62]*De Genesi ad litteram* 2.8.16 (CSEL 28:1, 44).
[63]*De Genesi ad litteram* 2.8.16 (CSEL 28:1, 44). Augustine articulates this view again in *De Genesi ad litteram* 4.24.41 (CSEL 28:1, 123-24).

for Augustine, trinitarian agency in creation necessitates angelic agency in creation as well, since the angels incessantly gaze on the Word of God (after, of course, their *own* creation). An analogy: Suppose a wizard creates rabbits by first picturing them in a crystal ball for a period of time, and then actually making them. Now, if the wizard has associates who also see the crystal ball, they will necessarily see the rabbits along with him before the rabbits actually exist.

But what does all this have to do with how Augustine reads Genesis 1? He argues in book 2 of his literal commentary that the phrase "and God said, 'let there be'" throughout Genesis 1 corresponds to creation in the Word; the phrase "and thus it was made" corresponds to creation in the angelic knowledge; and the phrase "and God made this and that" corresponds to the actual creation of the object.[64] Later, in book 4, he picks up this theme again, this time involving the text's reference to "morning and evening" as well:

> And so in this account of the creation of things the day is not to be understood as the form of the actual work, nor the evening as its termination and the morning as the start of another work. . . . Instead, that "day which God has made" is itself repeated through his works, not in a bodily circular motion but in spiritual knowledge, when that blessed company of angels before anything else contemplates in the Word of God that about which God says *Let it be made*; and in consequence this is first made in their own angelic knowledge when the text says, *And thus it was made*, and only after that do they know the actual thing made in itself, which is signified by the making of the evening.[65]

Here Augustine explicitly denies a sequential interpretation in which the morning of a creation day marks the beginning of a particular creative work and the evening marks its termination. Rather, he

[64]*De Genesi ad litteram* 2.8.19 (CSEL 28:1, 45).

[65]*De Genesi ad litteram* 4.26.43 (CSEL 28:1, 125). The quotation here is a reference to Psalm 118:24, which Augustine has just cited in 4.25.42. Cf. his parallel discussion of angelic knowledge in *De civitate Dei* 11.29 (CSEL 40.1, 556-57).

maintains that these creation days are ordered according to "spiritual knowledge," that is, angelic knowledge.

In book 4 Augustine also develops this conception of the "morning" of each creation day, identifying it as angelic praise for the Truth in which they see each thing created.[66] He then spends considerable time extending his comments on the nature of angelic knowledge and how the sequential ordering of Genesis 1 relates to instantaneous creation. He uses different kinds of language to describe instantaneous creation—for instance, he will often speak of one "day" of creation repeated six or seven times. Nonetheless, he makes it clear that the structuring of this day into a six-or-seven-day pattern reflects the order of time but of causation,[67] and that "this sixfold or sevenfold repetition happened without any intervals or periods of time."[68] This one "day" repeated six or seven times is thus not a period of time at all, but rather "the harmonious fellowship and unity of the supercelestial angels and powers."[69] Thus, when Augustine speaks of the "days of creation" or the "repetition of the days," this language is not at odds with his affirmation of instantaneous creation.

Augustine summarizes his view toward the end of book 4, emphasizing again the pedagogical, as opposed to strictly literary or thematic, purpose of the depiction of an instantaneous creation in terms of a six-or-seven-day week:

> The one who made all things simultaneously together also made simultaneously these six or seven days, or rather this one day six or seven times repeated. So then, what need was there for the six days to be recounted so distinctly and methodically? It was for the sake of those who cannot arrive at an understanding of the text, "he created

[66]*De Genesi ad litteram* 4.26.43 (CSEL 28:1, 125).
[67]*De Genesi ad litteram* 5.5.15 (CSEL 28:1, 147).
[68]*De Genesi ad litteram* 5.3.6 (CSEL 28:1, 141).
[69]*De Genesi ad litteram* 5.4.10 (CSEL 28:1, 144).

all things together simultaneously," unless scripture accompanies them more slowly, step by step, to the goal to which it is leading them.[70]

Here Augustine asks: If creation was instantaneous, why is it depicted in six days, particularly in six days involving such detail and progression ("distinctly and methodically")? His answer—that "it was for the sake of those who cannot arrive at an understanding . . . unless Scripture accompanies them more slowly"—is drawn from his conviction that God's work of creation is ineffable and difficult, and therefore requires an act of accommodation to understand. But Augustine's appeal to Sirach 18:1 in this context ("he created all things together simultaneously") indicates that there are also textual factors involved in his exegesis (more on that in just a moment).

Although Augustine thinks everything in Genesis 1 was created simultaneously, this does not entail that he denies any differences in God's creative work throughout this chapter. He emphasizes, particularly throughout book 1 of his literal commentary, the difference between the creation event referenced in Genesis 1:1-2 and God's creative work throughout 1:3–2:3—the latter section is distinguished, among other things, by the creation days, and by the various repeated formulae throughout them, such as divine speech ("and God said, 'let there be'"), divine sight ("and God saw"), divine naming, and so forth. At one point, he even considers the notion that the earth and the various kinds of water on it were created *prior* to days one to six.[71] At another point, he stipulates that Genesis 1:1-2 refers to God's creation of "basic material," while the subsequent verses refer to God's forming and fashioning that material into its present form. After citing Wisdom 11:28 for support, Augustine declares:

It is considerations of this sort, you see, that have convinced me that it was this basic material that was indicated by those words which

[70]*De Genesi ad litteram* 4.33.52 (CSEL 28:1, 133).
[71]*De Genesi ad litteram* 1.13.27 (CSEL 18:1, 19-20).

> spiritual foresight adapted even to less quick-witted readers or listeners, and which say before coming to any count of days, *In the beginning God made heaven and earth*, and so on until it comes to *And God said*, so that from there on the order of things given form and shape might follow.[72]

Note again here Augustine's appreciation of the sensitivity with which biblical language functions, taking into view the intellectual capabilities of its audience.

Augustine's emphasis on functional creation throughout Genesis 1 bears certain resemblances to John Walton's view, although they differ in that Augustine sees functional creation only starting with Genesis 1:3.[73] Augustine maintains that material creation (Gen 1:1–Gen 1:2) does not precede functional creation (Gen 1:3–2:3) temporally, since all things were made simultaneously: "God the Creator did not first make formless material and later on form it, on second thoughts as it were."[74] Rather, Augustine regards material creation as the *source* of functional creation, since "that which something is made out of is still prior as its source, even if not in time."[75] In Augustine's view, the entirety of Genesis 1 happened simultaneously, but it nonetheless made sense to record the source first and the result second: "God made them simultaneously, both the material which he formed and the things into which he formed it, and since both had to be mentioned by scripture and both could not be mentioned simultaneously, can anybody doubt that what something was made out of had rightly to be mentioned before what was made out of it?"[76] It is striking that Augustine could entertain the notion of functional creation in a sequence of narration as detailed as Genesis 1:3–2:3. At the same time,

[72]*De Genesi ad litteram* 1.14.28 (CSEL 18:1, 20-21).

[73]John H. Walton, *The Lost World of Genesis One: Ancient Cosmology and the Origins Debate* (Downers Grove, IL: IVP Academic, 2009).

[74]*De Genesi ad litteram* 1.15.29 (CSEL 18:1, 21).

[75]*De Genesi ad litteram* 1.15.29 (CSEL 18:1, 21).

[76]*De Genesi ad litteram* 1.15.29 (CSEL 18:1, 21-22).

because he regarded all of Genesis 1 as depicting an act of instantaneous creation, functional creation and material creation were ultimately not separated from each other in his thought.

TEXTUAL REASONS WHY AUGUSTINE HELD THIS VIEW

Although Augustine was alert to broader philosophical issues in his context, his interpretation of Genesis 1 was ultimately rooted in certain exegetical concerns. To be sure, his theological instincts were doubtless at play as well. For instance, he seemed to think that instantaneous creation accorded better with divine omnipotence. At one point, when he reminds his readers that the depiction of the creation of the sea and earth "is not to be taken as involving intervals of time," he stipulates the underlying rationale for this restriction as "in case the inexpressible ease with which God works should be limited by some kind of slowness."[77] Nonetheless, the greater portion of Augustine's rationale for instantaneous creation was rooted in various features of the text. Here we identify several of the most important factors.

1. The "light before the luminaries" conundrum. First, Augustine wrestled with the nature of the light in days one to three before the creation of the luminaries on day four. Throughout books 1 and 2 of the literal commentary, for instance, Augustine openly wondered how to relate the light of day one with the "luminous bodies" of day four. Early on, he raises the possibility that the pre-solar light of day one is spiritual light, not corporeal, but he does not provide an answer.[78] Throughout these books, he offers numerous possible answers, such as that the earlier light was illuminating "higher regions far from the earth," or the claim evidently made by another commentator known to Augustine (but not to us) that day one deals with the essence of

[77] *De Genesi ad litteram liber unus imperfectus* 11.34 (CSEL 28:1, 483).
[78] *De Genesi ad litteram* 1.3.7 (CSEL 28:1, 7).

light, while day four deals with the organization of light into days.[79] But Augustine recognizes that all of these proposals generate difficulties. Noting the phrase "let them be for signs and for seasons, and for days and years" in Genesis 1:14, Augustine asks, "Who can fail to see how problematic is their implication that times began on the fourth day, as though the preceding three days could have passed without time?"[80] This problem greatly vexed Augustine. He calls it a "mystery" and a "secret."[81] In one passage, he observes that Genesis 1 does not inform us where light went in the evenings of days one to three, and therefore we cannot know.[82] He regards it as implausible and artificial that God would arbitrarily suspend light before day four: "It can scarcely be supposed, after all, that it was put out so that nocturnal darkness might follow, and then lit again so that morning might be made, before the sun took on this task, which as the same text testifies was made to begin on the fourth day."[83] Ultimately, he seems to arrive at only uncertainty: "What mind, therefore, is capable of penetrating the mystery of how those three days passed before times began, times which are said to have begun on the fourth day—or of whether indeed those days passed at all?"[84]

In book 4 of the literal commentary, Augustine finds himself dragged back into the same issue: "We find ourselves slipping back into the same problem . . . and having to ask how light could circulate to produce the alternations of day and night, not only before the lamps of heaven but even before heaven itself, called the solid structure, was made."[85] Augustine recalls his earlier position that this light refers to the spiritual creation, but now seems uncertain: "At this

[79]*De Genesi ad litteram* 1.11.23 (CSEL 28:1, 17).
[80]*De Genesi ad litteram* 2.14.28 (CSEL 28:1, 53-54). See also Augustine's treatment of this problem in *De civitate Dei* 11.7 (CSEL 40.1, 520-21).
[81]*De Genesi ad litteram* 2.14.28 (CSEL 28:1, 54).
[82]*De Genesi ad litteram* 1.11.23 (CSEL 28:1, 17).
[83]*De Genesi ad litteram* 1.11.23 (CSEL 28:1, 17).
[84]*De Genesi ad litteram* 2.14.28 (CSEL 28:1, 54).
[85]*De Genesi ad litteram* 4.21.38 (CSEL 28:1, 120).

stage we have been forcibly reminded by our reflection on the seventh day that the easy and sensible thing to do is to admit our ignorance about what is so remote from any experience of ours, and say that we simply do not know how that light which was called day, if it means bodily light, affected the alternations of light and day."[86] Augustine's willingness here in book 4 to retract the interpretation advanced in book 1 now that the complication of divine rest has come to the foreground speaks once again to the hermeneutical flexibility he finds appropriate outside the rule of faith. Further, his language for the topic at hand ("so remote from any experience of ours") underscores his view of creation as a unique, mysterious topic.[87]

Later in his commentary he will return to the view that the light in Genesis 1:3 pertains to angels,[88] and this is the view he advocates in the *Confessions* as well: "At the beginning of your creation you said *Let there be light; and the light began.* I think these words are properly to be understood to refer to the spiritual creation."[89] Here Augustine envisions Genesis 1:3 not as referring to the coming into existence of the angelic realm, but as its passing from its formless state into an enlightened state. The angelic kingdom received light, Augustine maintains, not in an automatic manner by virtue of its mere existence, but by "fixing its gaze upon you and clinging to you, the Light which shone upon it."[90] In this context, Augustine associates the light bestowed on angels with beatitude, and he emphasizes that God is the source of both the angels' existence and their light/beatitude, since God alone is simple and immutable.[91] In *The City of God*, Augustine

[86]*De Genesi ad litteram* 4.21.38 (CSEL 28:1, 120).

[87]Another example would be Augustine's difficulty interpreting the phrase "heaven and earth" in Genesis 1:1. He spends a great deal of energy laboring to understand the nature (physical or spiritual) of the "heaven" in view here, and goes back and forth throughout his writings. E.g., cf. his deliberations in *Confessiones* 12.17 (CSEL 33, 325-27).

[88]*De Genesi ad litteram* 5.19.37 (CSEL 28:1, 161).

[89]*Confessiones* 13.3 (CSEL 33, 347).

[90]*Confessiones* 13.3 (CSEL 33, 347).

[91]*Confessiones* 13.3 (CSEL 33, 347).

again identifies the angels with the light of Genesis 1, confessing that he is unable to believe that they are simply omitted from the account, and reasoning that they certainly cannot come after the creation week, since Job 38:7 declares they were praising God when he made the stars.[92] This led Augustine to differentiate the days of creation from ordinary days, since "our ordinary days have no evening but by the setting, and no morning but by the rising, of the sun; but the first three days of all were passed without the sun."[93] Augustine therefore maintained here, as elsewhere, that the "evening and morning" of the creation days do not refer to successive times, but to creaturely knowledge and praise, respectively.[94]

2. Relating Genesis 2:4-6 to Genesis 1:1–2:3. A second textual difficulty that weighed on Augustine was the challenge of relating Genesis 2:4-6 to the creation week of Genesis, particularly the different usage of the word "day" in 2:4 and the apparent dischronology introduced in 2:5 ("when no shrub had yet appeared"). Augustine had previously observed that the word "day" is used in different ways in Genesis 1—sometimes to refer to sunrise to sunset, other times to refer to a twenty-four-hour period of time "in the same way as when we say, for example, there are thirty days in a month."[95] Now Augustine recognizes a still further usage of "day" in Genesis 2:4. He devotes the entirety of book 5 of his literal commentary to the claim that the challenges involved in interpreting Genesis 2:4-6 provide a "more definite confirmation" of instantaneous creation, which he had argued for in the previous book more strictly from Genesis 1.[96] Since Genesis 2:4 references "heaven and earth" as well as "the greenery of the field" (which was created on the third day in Genesis 1), Augustine regards it as "beyond a shadow of doubt" and "limpidly clear" that the

[92]*De civitate Dei* 11.9 (CSEL 40.1, 522-25).
[93]*De civitate Dei* 11.7 (CSEL 40.1, 520).
[94]*De civitate Dei* 11.7 (CSEL 40.1, 521).
[95]*De Genesi ad litteram liber unus imperfectus* 7.28 (CSEL 28:1, 478).
[96]*De Genesi ad litteram* 5.1.1 (CSEL 28:1, 137).

days do not follow each other in successive temporal duration.[97] The very reason that "the greenery of the field" is added to the reference to "heaven and earth" in Genesis 2:4, Augustine holds, is to "drive home the point" that we must conceptualize the day of creation as "certainly not such as the one we are familiar with here."[98] Thus, if anyone should have escaped Genesis 1 still thinking of the days as twenty-four-hour periods of time, this error will be corrected once they arrive at the language of Genesis 2:4-6: "If readers should happen to understand here one of our ordinary kind of days, they would be corrected when they recalled that God said the earth should produce the greenery of the field before this ordinary solar day."[99]

Part of what may have motivated Augustine to give so much attention to Genesis 2:4-6 is the worry that his view of instantaneous creation in Genesis 1 may draw too heavily on Sirach 18:1 in the Old Latin version ("he who remains for eternity created all things at once"). So now Augustine is able to appeal to the textual proximity of Genesis 2:4-6: "Now we get evidence in support, not from another book of holy Scripture that God created all things simultaneously, but from next door neighbor's (*vicina*) testimony on the page following this whole matter, which gives us a hint with the words, *when the day was made God made heaven and earth and all the greenery of the field*."[100] Augustine proceeds to clarify that when he speaks of the one day of creation repeated six or seven times, he is not thinking that this repetition required a temporal duration: "This sixfold or sevenfold repetition happened without any intervals or periods of time."[101]

Augustine regards the reference to the "spring" in Genesis 2:6 as referring to what God began to do after the initial creative act, involving the development of the famous "seminal reasons" (*rationes*

[97] *De Genesi ad litteram* 5.1.3 (CSEL 28:1, 138).
[98] *De Genesi ad litteram* 5.2.4 (CSEL 28:1, 139-40).
[99] *De Genesi ad litteram* 5.3.6 (CSEL 28:1, 141).
[100] *De Genesi ad litteram* 5.3.6 (CSEL 28:1, 141).
[101] *De Genesi ad litteram* 5.3.6 (CSEL 28:1, 141).

seminales) that have been implanted into creation at the creative moment.[102] Thus, Augustine understands Genesis 2:4-5 as a recapitulation of the entire instantaneous creative work, with Genesis 2:6 then narrating God's subsequent work. He therefore conceives of Genesis 2:4-5 as a significant turning point or hinge in the biblical narrative. As he puts it, "Between those works of God from which he rested on the seventh day, and these which he is working on until now, scripture inserted a kind of joint in her narrative."[103] Augustine thus divides God's creative work into two basic categories: the instantaneous initial creative work recounted in Genesis 1:1–2:3, and God's subsequent work over intervals of time beginning in Genesis 2:6. He repeatedly emphasizes the distinctness of these two kinds of work: "God worked in one way with all creatures at their establishment . . . and in another at their management and regulation."[104] Augustine carries this distinction into the rest of his theology, including his preaching, frequently invoking Jesus' reference to the Father's "working until now" in John 5:17.[105]

3. God's "rest" on the seventh day. One reason why Augustine emphasizes the distinctness of God's work in the initial creation act from his subsequent work is the text's account of God's rest on the seventh day. Divine rest plays a significant role throughout Augustine's theology, and his pastoral vision as well. In his preaching, for instance, he exhorted his listeners regarding the superiority of rest over work, since rest is sanctified in Genesis 1 and work is not, and since work ceases with the six days but the seventh day of rest continues on.[106] Ultimately, the motif of rest became a dominant symbol for his conception of heavenly beatitude, typified by Mary sitting at

---

[102] *De Genesi ad litteram* 5.7.20 (CSEL 28:1, 150).

[103] *De Genesi ad litteram* 5.11.27 (CSEL 28:1, 155).

[104] *De Genesi ad litteram* 5.11.27 (CSEL 28:1, 154-55).

[105] For his immediate discussion of this text, see *De Genesi ad litteram* 4.11.21 (CSEL 28:1, 107-8).

[106] Sermon 8.17, in *Sermons 1–19: On the Old Testament*, ed, John E. Rotelle, trans. Edmund Hill, The Works of Saint Augustine: A Translation for the 21st Century (Brooklyn, NY: New City Press, 1990), 251; Sermon 125.4, in *Sermons 94A–147A*, 255-56.

the feet of Jesus in Luke 10:39.[107] When he addresses the nature of the divine rest of Genesis 1 in his literal commentary, he appears profoundly energized, and he devotes the bulk of book 4 of his literal commentary to this topic.[108] It is axiomatic to him to assert that "God did not delight in some kind of temporal period of rest after hard toil."[109] It is "the height of folly" to entertain any conception that God's "emotional tensions over this business were first relieved," since God's faculties are "incomparably and inexpressibly sufficient for the creation of things."[110] But if God cannot get tired, why is he said to rest? Augustine claims that this language of God's rest must be taken analogically, in the sense that it refers to God's creation of the rest that creatures enjoy. He then develops this linguistic claim by referencing a number of other biblical passages in which God is said to be affected in some manner, and yet the language actually refers to *creatures* being changed.[111]

But Augustine proceeds to inquire whether there is any sense in which God himself can be spoken of as resting. He considers a number of possible ways in which this might be affirmed, such as the claim that divine rest means that God stopped creating new kinds of creatures,[112] or divine rest means God's self-sufficient happiness.[113] A central feature of his thought on this question is the fact that the seventh day has a morning but no evening. Stemming from this, Augustine claims that the seventh day is not a period of time that God creates, since God made all things in one day (indeed, in one

[107]*De Genesi ad litteram* 4.14.25 (CSEL 28:1, 111). I explore how this Mary-Martha typology developed through subsequent patristic and medieval theology in my *Anselm's Search for Joy: An Interpretation of the* Proslogion (Washington, DC: Catholic University of America Press, 2020).

[108]*De Genesi ad litteram* 4.8.15-20.37 (CSEL 28:1, 103-20).

[109]*De Genesi ad litteram* 4.14.25 (CSEL 28:1, 111).

[110]*De Genesi ad litteram* 4.8.16 (CSEL 28:1, 104).

[111]*De Genesi ad litteram* 4.9.17-19 (CSEL 28:1, 104-6). Here he references passages such as God's knowledge of Abraham in Genesis 22:12, the Holy Spirit's being grieved despite possessing perpetual bliss in Ephesians 4:30, and God's knowledge in Galatians 4:9 and 1 Peter 1:20.

[112]*De Genesi ad litteram* 4.12.22 (CSEL 28:1, 108).

[113]*De Genesi ad litteram* 4.16.27 (CSEL 28:1, 112-13).

instantaneous moment). As he puts it, "The seventh day is not a creature on its own, but the same one coming round seven times, the one which was fashioned when God called the light day and the darkness night."[114] He speaks of the repetition of these "many so-called days" on the angelic intelligence, deducing that "there was no need to create the seventh day, because it was made, naturally, by the seventh repetition of this one day."[115] Augustine then reasons that the presence of day seven suggests that the first six days of creation unfolded "in a manner quite beyond what we are used to in our experience, with the original fashioning of things, so that in them neither evening nor morning, neither light itself nor darkness, presented the same alternations as they do in these days through the circling round of the sun."[116] Thus, for Augustine, the presentation of divine rest at the end of Genesis 1 provides a further confirmation to that already supplied by Genesis 2:4-6 that the days of creation in Genesis 1 are mysterious and beyond our ordinary experience.

These three features of the creation account—the light before luminaries issue, Genesis 2:4-6, and divine rest—are perhaps the primary textual considerations that led Augustine to emphasize that the days of creation are not ordinary days. But they are not the only ones. One other struggle that possessed Augustine was the challenge of fitting the events narrated into the timeframe of twenty-four hours for each day. After noting the multiplication of animals in Genesis 1:22, for instance, he appeals:

> Here, surely, anyone slow on the uptake should finally wake up to understanding what sort of days are being counted here. . . . How could [animals] in one day both conceive and carry in the womb and hatch what they have laid and rear them and fill the waters of the sea

---

[114]*De Genesi ad litteram* 4.20.37 (CSEL 28:1, 120).

[115]*De Genesi ad litteram* 4.20.37 (CSEL 28:1, 119); Hill includes the metaphor of a "screen" in his translation of this sentence to convey the nature of the presentation of these days to the angels.

[116]*De Genesi ad litteram* 4.18.33 (CSEL 28:1, 116).

> and multiply upon the earth? It does, after all, add *and thus it was made* before the evening comes.[117]

Interestingly, Augustine makes this argument with respect to day five, not day six (as is more common among those arguing against twenty-four-hour days today). The sharpness of Augustine's language on this point, calling for "slow" interpreters to "wake up," must be understood in relation to Augustine's apologetic burden, informed by his own long captivity to Manichaeism. But it also suggests a striking level of confidence, even indignation, in his protest against twenty-four-hour creation days.

DOES AUGUSTINE CARE ABOUT AUTHORIAL INTENT?

At this point it may be useful to inquire why we should care about Augustine's exegesis of Scripture in the first place. For many modern readers of the Bible, it is not at all obvious that we should. Indeed, many of the views that we have surveyed in this chapter will likely seem bizarre to modern interpreters. And this has been primarily working with the literal commentary; if we turn to the allegorical commentary, we could stack up further Augustinian oddities such as the following:

- the solid structure of the dome of Genesis 1:6 separates the "basic bodily material of visible things" from "the basic non-bodily material of things invisible";[118]
- the greenery of the field in Genesis 2:5 indicates the invisible creation, such as the soul;[119]
- the spring coming up from the earth in Genesis 2:6 refers to "the flood of truth drenching the soul before sin";[120]

---

[117] *De Genesi ad litteram liber unus imperfectus* 15.51 (CSEL 28:1, 495).
[118] *De Genesi contra Manichaeos* 1.11.17 (CSEL 91, 83).
[119] *De Genesi contra Manichaeos* 2.3.4 (CSEL 91, 122).
[120] *De Genesi contra Manichaeos* 2.6.7 (CSEL 91, 126).

- the four rivers in Eden in Genesis 2:10-14 represent the four spiritual virtues of prudence, fortitude, temperance, and justice;[121]
- the angel's flaming sword in Genesis 3:24 expresses "temporal punishments and pains, since time goes whirling and spinning along."[122]

The fact that Augustine ultimately shifts from the allegorical method that underpins these interpretations does not necessarily remove him from suspicion in the eyes of many modern interpreters. That he generates such ideas at any time, in any type of commentary, raises eyebrows. Besides, even in the literal commentary we encounter claims such as that the water of Genesis 1:2 is a symbol of all material reality;[123] or that Adam was put to sleep in Genesis 2 so that his mind could "through ecstasy become as it were a member of angelic court;"[124] or all that we have surveyed about angelic knowledge in Genesis 1. What are we to say to those who reject Augustine's hermeneutic as haphazard and absurd?

We cannot fully respond to this difficulty here. A full defense would require us to venture into a larger discussion of patristic hermeneutics, for in the ways in which Augustine is eccentric to us, he was rarely eccentric in his own context.[125] But we can at least address one of the most prominent questions about his biblical exegesis, and one that is relevant to our reflections in this chapter: Does Augustine care about authorial intent, or does he use theology as the final determiner of biblical meaning?

This is a frequent charge against Augustine—that he allows his theology to dictate his exegesis. Blake Dutton, for instance, claims that

---

[121] *De Genesi contra Manichaeos* 2.10.14 (CSEL 91, 134-35).

[122] *De Genesi contra Manichaeos* 2.23.35 (CSEL 91, 158).

[123] *De Genesi ad litteram* 1.5.11 (CSEL 28:1, 9-10).

[124] *De Genesi ad litteram* 9.19.36 (CSEL 28:1, 294).

[125] For a discussion of patristic exegesis, see Graves, *Inspiration and Interpretation of Scripture*; John J. O'Keefe and R. R. Reno, *Sanctified Vision: An Introduction to Early Christian Interpretation of the Bible* (Baltimore: Johns Hopkins University Press, 2005); and Mark Sheridan, *Language for God in the Patristic Tradition: Wrestling with Biblical Anthropomorphism* (Downers Grove, IL: IVP Academic, 2015).

what makes a biblical interpretation acceptable for Augustine is that "it imputes to the text what is true rather than that it correctly explicates the meaning of the author."[126] Such a charge is not incomprehensible. We have seen, for example, how Augustine uses the rule of faith as a hermeneutical guideline. At the same time, Augustine would not have pitted biblical meaning and theological truth against one another, since he understood the rule of faith to be a faithful summary of biblical revelation. As it turns out, while Augustine did not display the rigorous interest of Jerome in issues like original languages and canonicity, authorial intended meaning was nonetheless crucial to him in interpreting the Bible, and his actual practice was more complicated than some caricatures will allow.[127]

One can see something of the complexity of Augustine's hermeneutics in his treatment of the meaning of the words "in the beginning" in Genesis 1:1. In *The City of God* he reaffirms that the angels are the "light" of Genesis 1:3. But he acknowledges an openness to the opposing view that this light is material light, provided that one takes the words "in the beginning" in Genesis 1:1 as meaning that God made all things by his Wisdom or Word, rather than as establishing that nothing happened before that moment. Augustine sees this alternative trinitarian interpretation of Genesis 1:1 as possible in light of Christ's claim to have spoken the truth "since the beginning" in John 8:25, and says that if someone takes this view, "I will not contest the point, chiefly because it gives me the liveliest satisfaction to find the Trinity celebrated in the very beginning of the book of Genesis."[128] Now, all of this might seem to suggest that Augustine is happy to derive "the right doctrine from the wrong text," so to speak. In other

[126]Blake D. Dutton, "The Privacy of the Mind and Fully Approvable Reading of Scripture," in *Augustine's Confessions: Philosophy in Autobiography*, ed. William E. Mann (Oxford: Oxford University Press, 2014), 155.

[127]On the different approaches to biblical interpretation in Augustine and Jerome, see the discussion in Gerald Bray, *Augustine on the Christian Life: Transformed By the Power of God*, Theologians on the Christian Life (Wheaton, IL: Crossway, 2015), 97-98.

[128]*De civitate Dei* 11.32 (CSEL 40.1, 560).

words, since Augustine wants to affirm the Trinity, he is content to see it in Genesis 1:1, regardless of the passage's intended meaning.

But not so fast. In the first place, we know from his other works that Augustine himself finds a christological interpretation of Genesis 1:1 plausible. In his *Confessions*, for instance, he parses the terms "in the beginning" (*in principio*) as "in your Word, in your Son, your strength, your wisdom, your truth " (*in verbo tuo, in filio tuo, in virtute tua, in sapientia tua, in veritate tua*), marveling at how this Wisdom can pierce and warm his heart.[129] In this context, Augustine reflects on the significance of God's creating the world through his Word, whom he identifies as God's Wisdom and Light and Truth.[130] Augustine then uses these considerations to establish that because God's Word is the source of all wisdom, "He is therefore the Beginning, the abiding Principle."[131] Augustine's openness to a christological interpretation of "in the beginning" in *The City of God* is rooted in this theological vision of the Son of God's role in creation.

Then, in *The City of God*, Augustine proceeds to ground his openness to different views of Genesis 1:1 in further textual considerations involving both Genesis 1 and Psalm 104. In particular, Augustine is intrigued by the reference to the Spirit in Genesis 1:2, which he regards as strengthening the plausibility of a christological referent in Genesis 1:1.[132] Since Augustine develops his openness to an alternative interpretation by a consideration of features *in the text*, it is not quite fair to portray him as saying, "The author meant one thing, but we can deviate from it so long as we stay within the rule of faith." Rather, Augustine seems to think the text itself opens the door to these interpretive possibilities. This is why he concludes by grounding his openness to different interpretations in the text's profundity rather

[129]*Confessiones* 11.9 (CSEL 33, 288), my translation.
[130]*Confessiones* 11.6-7 (CSEL 33, 285-87).
[131]*Confessiones* 11.8 (CSEL 33, 287).
[132]*De civitate Dei* 11.32 (CSEL 40.1, 560-61).

than the rule of faith: "It is so profound a passage, that it may well suggest, for the exercise of the reader's tact, many opinions, and none of them widely departing from the rule of faith."[133] Such a hermeneutical practice is certainly not beyond criticism, but it is not easy to say that it reflects a lack of care for the author's meaning. It would be more accurate to say that Augustine complicates authorial meaning than simply departs from it.

Moreover, we have observed Augustine elsewhere earnestly desiring that Moses were still alive, so that he might lay ahold of him and "beg and beseech" him to explain these very words in Genesis 1:1: "I would be all ears to catch the sounds that fell from his lips."[134] Throughout his exegesis of this verse, Augustine frequently speaks of "the meaning which Moses had in mind,"[135] making clear that this is his ultimate interest. After referencing two different interpretations of what is signified by the words "heaven and earth" in Genesis 1:1, Augustine admits, "I see that either meaning could be the true one, whichever of the two he may have meant. But it is not so clear to me which of them he in fact had in mind."[136] In this case and many others, Augustine's openness to multiple interpretations does not result from a lack of interest in Moses' intention but his uncertainty as to what it was. It is instructive that in such situations Augustine is willing, sometimes more so than contemporary interpreters of the Bible, to navigate in terms of probabilities. He is comfortable with not only ignorance, but with relative shades of confidence. He is willing to say not only, "I don't know," but also, "Here are three possible views; I think number two is right, but it might be number three."

A further complicating factor is that Augustine thinks that on some occasions Moses himself may have had multiple meanings in mind:

---

[133]*De civitate Dei* 11.32 (CSEL 40.1, 561).
[134]*Confessiones* 11.3 (CSEL 33, 283).
[135]*Confessiones* 12.24 (CSEL 33, 334).
[136]*Confessiones* 12.24 (CSEL 33, 334).

> I hear people say "Moses meant this" or "Moses meant that." I think it more truly religious to say "Why should he not have had both meanings in mind, if both are true? And if others see in the same words a third, or a fourth, or any number of true meanings, why should we not believe that Moses saw them all? There is only one God, who caused Moses to write the Holy Scriptures in the way best suited to the minds of great numbers of men who would all see truths in them, though not the same truths in each case.[137]

Here it is evident that Augustine's willingness to affirm multiple valid senses of the truth of Scripture should not be set over and against an interest in authorial meaning, since he thinks that multiple meanings may *be* the intended authorial meaning. Augustine grounds this claim in the divine inspiration of Scripture—he thinks the way God worked through Moses reflected a sensitivity to his large and diverse readership. At the same time, Augustine's affirmation of multiple meanings in Scripture was not unrestrained, and as he continues he sets strictures against the abuse of this principle: "If he had only one meaning in mind, let us admit that it must transcend all others."[138]

Ultimately, while the rule of faith was vital for Augustine's biblical interpretation, he never set this at odds with the intended meaning of the author as a hermeneutical consideration. For instance, at the end of the first book of the literal commentary, Augustine writes:

> When we read in the divine books such a vast array of true meanings, which can be extracted from a few words, and which are backed by sound Catholic faith, we should pick above all the one which can certainly be shown to have been held by the author we are reading; while if this is hidden from us, then surely one which the scriptural context does not rule out and which is agreeable to sound faith; but if even scriptural context cannot be worked out and assessed, then at least only one which sound faith prescribes.[139]

[137] *Confessiones* 12.31 (CSEL 33, 343).
[138] *Confessiones* 12.32 (CSEL 33, 343).
[139] *De Genesi ad litteram* 1.21.41 (CSEL 28:1, 31).

Here Augustine articulates a hierarchy of priorities in making exegetical conclusions, and identifies the intended meaning of the author as chief among them. Moreover, "sound faith" is invoked by itself only if it cannot be used in conjunction with "scriptural context." Thus, not only does Augustine refrain from pitting authorial intent and the rule of faith against one another, but in a crucial sense he prioritizes the former over the latter.[140]

Augustine's biblical exegesis is quite complicated, and we have only scratched the surface of the issues that could be explored here. But at the very least, the recognition of Augustine's interest in authorial intent, and the complexity of his hermeneutics, ought to discourage hastiness or dismissiveness in our evaluations. It is easy for modern interpreters to regard certain habits of patristic exegesis as hopelessly outdated, but recall our travel metaphor from the introduction: engaging Augustine's exegesis is like visiting a foreign country that has a very different culture than our own. In such a setting, we should not be hasty in our judgments, but take a long time to listen and observe. Similarly, in our venture into Augustine's treatment of Genesis 1, it will be helpful to patiently ask, "How did this make sense to him?" before launching criticisms.

Ultimately, the sheer stature of Augustine's mind and faith should encourage us to focus less on the shortcomings of his efforts and more on the difficulty of the task at hand. If mighty Augustine could struggle with comprehending the meaning of Genesis 1, surely there is space for us to do the same.

AUGUSTINE'S VIEW IN HISTORICAL CONTEXT

From a different angle, some may wish to evade the implications of Augustine's views on Genesis 1 by isolating him. *That's just Augustine*,

[140]Bray, *Augustine on the Christian Life*, 100-107, has a helpful discussion of these passages, interpreting Augustine on intended meaning in terms of his distinction between "things" and "signs."

we might admit with a shrug. *He said lots of strange things*. But with respect to his views of Genesis 1, Augustine was neither eccentric in his time nor insignificant in his influence upon subsequent generations. For instance, Augustine's effort to draw Genesis 1 into contact with the whole of divine revelation, reading it in an explicitly trinitarian and christocentric framework, was a common feature of patristic exegesis. As Paul Blowers puts it, "Genesis 1 was already, for patristic exegetes, a 'tableau' or montage of the whole divine economy, and pointed ahead to the mystery of salvation and the appearance of the 'new creation.'"[141] Blowers emphasizes that Augustine's nuanced approach to the biblical creation account shares fundamental continuities with broader practice among the church fathers:

> The consensus of early Christian theologians . . . shared Origen's conviction that divine revelation was inexhaustible in its nuances and horizons of meaning. . . . While Origen had asserted that the Spirit of God providentially problematized sacred texts to exercise and instruct rational creatures, subsequent patristic commentators similarly engaged Scripture's complexity as altogether salutary, in the manner of a good dramatic script. Their analytical interpretation of the Hexaemeron . . . is a classic case in point. Augustine—in his *ad litteram* exegetical approach, no less—seems to relish precisely in the text's intricacy, recognizing that the Hexaemeron is a thick narrative, a cosmogonic montage, presaging the whole developmental history of the cosmos under the wise providence of the Trinity. While committed to the simplicity and coherence of the church's Rule of Faith, Augustine shows profound patience with the density and complexity of Genesis 1 as well as resistance to facile reading of the text when its constituent details may, in light of the Bible as a whole, carry multiple plausible meanings.[142]

[141]Paul M. Blowers, *Drama of the Divine Economy: Creator and Creation in Early Christian Theology and Piety*, Oxford Early Christian Studies (Oxford: Oxford University Press, 2012), 8.
[142]Blowers, *Drama of the Divine Economy*, 377.

With regard to the "days" of Genesis 1 specifically, Augustine was not the first to interpret the days of creation as different than ordinary days—earlier comparable "non-literal" readings could be found, for instance, in Clement of Alexandria, Origen, Didymus the Blind, and Athanasius.[143] Moreover, Augustine's view, particularly as later qualified by Gregory the Great and then propagated by both Isidore of Seville and Bede, exerted significant influence on subsequent interpretation of Genesis 1, particularly in the medieval church.[144] Andrew Brown calls it "the defining statement with which every medieval and Renaissance commentator on Gen. 1:1–2:3 would wrestle."[145] Much of the discussion in the medieval church was over how to relate the creation of angels to the creation week of Genesis 1. Augustine's influence can be seen on all sides of this debate, but especially among those who, because of the challenge of relating the creation of angels and Genesis 1, saw this text as something other than a comprehensive, sequential account of creation.[146] The notion of instantaneous creation, with its concomitant affirmation of the days as a literary framework, remained widespread as well. In the eleventh century, for

---

[143]See Andrew J. Brown, *The Days of Creation: A History of Christian Interpretation of Genesis 1:1–2:3*, History of Biblical Interpretation (Blandford Forum, UK: Deo, 2014), 26-31.

[144]Gregory synthesized Augustine's views with a broader notion of sequence, applying instantaneous creation to the initial creation of things in their substance but allowing sequence for the creation of their particular final forms. See Brown, *The Days of Creation*, 57-59, 62-64, 102.

[145]Brown, *The Days of Creation*, 53.

[146]Many medieval theologians believed angels were created prior to the material universe, finding support for this view in the depiction of angels singing for joy at the creation of the stars in Job 38:7 and (in some cases) by identifying angels with the wisdom created before the world in Proverbs and Sirach. Others identified angels with the luminaries of day four. Peter Lombard, however, argued that angels were created before the creation week of Genesis 1 but not prior to creation itself. Instead, Lombard applied Augustine's idea of instantaneous creation to both angels and the four basic elements of creation, correlating the timing of their creation with the events in view in Genesis 1:1. In fact, instead of identifying angels with the pre-world wisdom of Wisdom literature, Lombard identified angels with the "heavens" of Genesis 1:1. This stood in contrast to the view of Aquinas that the "heavens" of Genesis 1:1 included what we would call "intermediate heaven" (where deceased believers go before final resurrection). For many other medieval theologians, "heavens" in Genesis 1:1 simply meant the stars and bodies "above" the earth. For a helpful overview of this whole discussion, see Marcia L. Colish, *Medieval Foundations of the Western Intellectual Tradition 400–1400*, Yale Intellectual History of the West (New Haven, CT: Yale University Press, 1997), 274-301.

instance, Anselm, though he does not commit himself to any particular view, referred to the doctrine of instantaneous creation as the "prevailing opinion" of his time. Anselm is speaking in this context of human beings and angels being made at the same time; this is drawn from his earlier reference to the view that "the entire creation was made all at once, and the 'days' by which Moses seems to indicate that the world was not made all at once must be understood as something other than the days that we experience in our own lives."[147]

How does Augustine's interpretation of Genesis 1 square with modern views? Obviously, the content of Augustine's interpretation does not exactly correspond to any contemporary category of interpretation, whether the day-age, analogical day, intermittent day, gap theory, cosmic temple, or framework view. But its overall way of understanding the text is closest to a *literary* or *framework* interpretation of the days, with a concomitant emphasis on accommodation.[148] To be sure, his view has some of its own distinguishing eccentricities, such as the emphasis on instantaneous creation (it is an irony that while many modern readers regard one week as too brief a time in which to fit the creation of the universe, Augustine felt it was too long).[149] But Augustine does regard the depiction of God accomplishing his work of creation in six days as a literary framework, accommodated to the original hearers' understanding, and not concerned with establishing strict sequence or timing. Moreover, though he does not emphasize it as much as some modern framework proponents, Augustine finds significance in the two triads in which days

---

[147]Anselm, *Cur Deus Homo* 1.18, in *Anselm: Basic Writings*, ed. and trans. Thomas Williams (Indianapolis: Hackett, 2007), 270.

[148]Henri Blocher, a modern framework proponent, lists Augustine as the first proponent of this view (what he terms the "literary" view) in his *In the Beginning: The Opening Chapters of Genesis* (Downers Grove, IL: InterVarsity Press, 1984), 49.

[149]E.g., in *De Genesi ad litteram* 1.10.19-20 (CSEL 28:1, 14), he repeatedly calls it "astonishing" to think that God would create using the passage of time, wondering (for instance) "why it should take such a long time to make light, until the space of a whole day had passed, and evening could be made."

one and four, two and five, and three and six are correlated (a common emphasis of framework views). Augustine mentions this in the context of his emphasis on the perfection of the number six, observing the fittingness of the first three days depicting the creation of three distinct places, and the next three of those things that occupy and move about in those places.[150]

CONCLUSION

Augustine's treatment of Genesis 1 will be disappointing to different kinds of people. It betrays both the evasive tendencies of higher-critical scholarship and the literalistic leanings of fundamentalism. Augustine takes the text too seriously, and submits to it too reverently, to be acceptable according to liberal canons of interpretation. But Augustine will also be an annoyance to those who suppose that premodern exegesis of Genesis 1 was a calm, unperturbed matter, and that the interpretation of the text is obvious apart from scientific challenges. Indeed, having surveyed Augustine's engagement with the biblical creation account, we are in a better position to appreciate Stanley Jaki's claim that "interpretations of Genesis 1 produced in the patristic period, in its golden phase or not, should appear at times not so much as gold, not even as glitter, but something as confusing as smoke can be."[151] We also have at least some context by which to appreciate Andrew Brown's assertion that "any way the modern reader reads this creation account is almost incapable of being truly new."[152]

Those who insist that a twenty-four-hour-day interpretation is the only Christian view unmolested by nineteenth-century discoveries should be checked by this legacy of Augustinian exegesis. Even if we ultimately conclude that the days are twenty-four-hour periods of time, we should nevertheless appreciate the sincerity of the struggle

---

[150]*De Genesi ad litteram* 4.2.6 (CSEL 28:1, 97-98).
[151]Stanley Jaki, *Genesis 1 Through the Ages*, 2nd ed. (Royal Oaks, MI: Real View Books, 1988), 97.
[152]Brown, *The Days of Creation*, 4.

that many devout interpreters—including many pre-Darwinian interpreters—have had with textual details such as the nature and sequence of light, the different meanings of day (Hebrew *yom*) throughout the passage, and the presentation of divine activity such as rest. If we can accept that different views in this area can be orthodox even when wrong, we can avoid the unwelcome implication that St. Augustine must be anathematized as a "liberal."

Beyond this, Augustine's exegesis may invite us toward patience before the complexity of Genesis 1. Andrew Louth, describing the church fathers' interpretation of Genesis, observed, "There is none of the anxiety one finds in fundamentalist readings of Scripture today."[153] By this he meant that the fathers were not threatened by the scientific thought of their day, and they were not afraid to acknowledge the limits of biblical claims. At the same time, the fathers treated the Bible reverently, with an emphasis on its authority and an appreciation for its depth of meaning. For them, the text "bears any amount of careful pondering."[154]

What if, rather than seeing Genesis 1 as a matter of either obviousness or outdatedness, we saw it instead as a great ocean or forest—big enough to bear our deepest ponderings and anxieties?

[153]Andrew Louth, "The Fathers on Genesis," in *The Book of Genesis: Composition, Reception, and Interpretation*, ed. Craig A. Evans, Joel N. Lohr, and David L. Peterson, Supplements to Vetus Testamentum 152 (Leiden: Brill, 2012), 577; cf. also Louth, "The Six Days of Creation According to the Greek Fathers," in *Reading Genesis After Darwin*, ed. Stephen C. Barton and David Wilkinson (Oxford: Oxford University Press, 2009), 39-55.

[154]Louth, "The Fathers on Genesis," 577.

CHAPTER FOUR

"IN PRAISE OF ASHES AND DUNG"

Augustine on Animal Death

If God were to make anything imperfect, which he would then himself bring to perfection, what would be reprehensible about such an idea?

De Genesi ad litteram 2.15.30

One of the most contested issues in contemporary evangelical debate about the doctrine of creation concerns the prospect of animal death, predation, pain, disease, and extinction before the human fall. Indeed, some claim it is *the* issue. David Snoke, for instance, observes that in his experience, "This is the fundamental issue of Bible interpretation caught up in the debate," and that the interpretation of Genesis 1 is actually "of secondary importance."[1] Ronald Osborn suggests that "the suffering of animals may be the most severe theodicy dilemma of all given the fundamental questions it raises about God's character as Creator."[2]

[1]David Snoke, *A Biblical Case for an Old Earth* (Grand Rapids: Baker Books, 2006), 48.

[2]Ronald Osborn, *Death Before the Fall: Biblical Literalism and the Problem of Animal Suffering* (Downers Grove, IL: IVP Academic, 2014), 19. Two other books that helpfully draw attention to the importance of this topic are Michael J. Murray, *Red in Tooth and Claw: Theism and the Problem*

Why should animal death, and the related conditions of animal life, be so prominent? Part of the answer may be that, for many people (including me), the question has an emotional component.[3] Part of it doubtless lies, as Osborn highlights, in its implication for theodicy and apologetics.[4] Even more basically, I would suggest that it is here that we begin to push through the more strictly hermeneutical issues of reading Genesis 1–3 into the deeper theological *consequences* of these different readings. Our position on animal death touches on deeper intuitions about the nature of createdness and fallenness, the character of the prehuman world, and the shape of the biblical narrative. Young-earth, old-earth, and (evangelical) evolutionary creationists all conceive of the Christian story as *creation-fall-redemption*—but their differences involve more than the mere time scale involved in each phase.[5]

It is often claimed that old-earth and evolutionary creationist views represent a capitulation to Darwinian and secular claims, in contrast to the settled consensus of pre-nineteenth-century Christianity. In other words, all Christians believed that animal death was bad until the claims of modern science put pressure on this conviction. Thus, one prominent young-earth creationist describes his view of creation as the same as that of "Jesus, the apostles, and *virtually all orthodox*

---

of Animal Suffering (Oxford: Oxford University Press, 2008); and Christopher Southgate, *The Groaning of Creation: God, Evolution, and the Problem of Evil* (Louisville, KY: Westminster John Knox, 2008).

[3]Charles Darwin himself wondered, and he was not the last to do so, "What advantage can there be in the sufferings of millions of the lower animals throughout almost endless time?" (quoted in John C. Munday, Jr., "Animal Pain: Beyond the Threshold?" in *Perspectives on an Evolving Creation*, ed. Keith B. Miller [Grand Rapids: Eerdmans, 2003], 450).

[4]It is often observed that although the advent of evolutionary theory did not create the challenge of theodicy, it has in certain respects altered its conditions and intensified it. Thus Gaymon Bennett, "Introduction: Evil and Evolution," in *The Evolution of Evil*, ed. Gaymon Bennett, Martinez J. Hewlett, Ted Peters, and Robert John Russell (Göttingen: Vandenhoeck & Ruprecht, 2008), 10.

[5]Here our focus will primarily be on differences between a young-earth creationist view of animal death, on the one hand, and old-earth/evolutionary views, on the other. For an illuminating discussion of the differences between old-earth creationist and evolutionary creationist views on this topic, see Kenneth Keathley, J. B. Stump, and Joe Aguirre, eds., *Old-Earth or Evolutionary Creation? Discussing Origins with Reasons to Believe and BioLogos*, BioLogos Books on Science and Christianity (Downers Grove, IL: IVP Academic, 2017), 68-84.

Christians prior to 1800," and claims that "the idea of millions of years of animal death, disease, violence, and extinction is utterly incompatible with the Bible's teaching *as well as orthodox Christian teaching for two thousand years*."[6]

In this chapter, however, we draw attention to greater complexity in the historical record by engaging Augustine's view of animal death. There are other theologians who could be engaged for the same purpose.[7] But Augustine is a particularly helpful voice to consider on this topic, since he regularly faced the Manichaean criticism of the alleged evils of the animal kingdom in relation to their dualistic cosmology. It is striking that Augustine did not attribute such phenomena to the human fall but vigorously defended the current state of the animal kingdom as a reflection of the goodness and wisdom of God. In fact, as we will see, Augustine devoted considerable energy to expounding the goodness of things such as worms, flies, ants, rats, fleas, frogs, larger predatory mammals, thorns and thistles, physical pain, and death itself.

What led Augustine to emphasize the goodness of insects and carnivores at such length and with such conviction? Developing an

[6]Ken Ham, "Rejoinder," in J. B. Stump, ed., *Four Views on Creation, Evolution, and Intelligent Design*, Counterpoints (Grand Rapids: Zondervan, 2017), 67, 68-69, both italics added. In the earlier book in the same series, the young-earth contributors appealed to the same problem: "The problem of evil in the world is already hard enough to explain without the addition of millions of years of animal suffering to round it out. What is the justification for all that animal pain?" See Paul Nelson and John Mark Reynolds, "Young-Earth Creationism," in *Three Views on Creation and Evolution*, ed. J. P. Moreland and John Mark Reynolds, Counterpoints (Grand Rapids: Zondervan, 1999), 47.

[7]Ambrose expounds God's wisdom in creating carnivorous animals like lions and bears, and dangerous animals like elephants and poisonous serpents, in his *Hexameron* 6.5.30-39 (Saint Ambrose, *Hexameron, Paradise, and Cain and Abel*, trans. John J. Savage, Fathers of the Church 42 [Washington, DC: Catholic University of America Press, 1961], 246-53). Much of Ambrose's material is drawn from Basil, who develops an arsenal of moral lessons for humanity from the animal kingdom, including from carnivorous animals, in his *Hexaemeron* 7.3-9.6 (Basil, *Letters and Selected Works*, ed. Philip Schaff and Henry Wace, trans. Blomfield Jackson, Nicene and Post-Nicene Fathers 8 [1895; repr., Peabody, MA: Hendrickson, 1994], 91-107). In the context of this effort, Basil explicitly defends the goodness of venomous and dangerous animals despite their threat to human life: "Let nobody accuse the Creator of having produced venomous animals, destroyers and enemies of our life. Else let them consider it a crime in the schoolmaster when he disciplines the restlessness of youth by the use of the rod and whip to maintain order" (*Hexaemeron* 9.5, in *Letters and Selected Works*, 105). I am grateful to my friend Joel Chopp for directing me to several of these references.

overview of his thought on this point may serve not only to hamper the young-earth appeal to historical continuity, but more deeply to probe the nature of createdness, and in so doing deepen and nuance Christian hope in the face of suffering.[8]

We will explore two Augustinian principles in turn: "temporal beauty," which will help us understand why Augustine thinks animal death is good; and "perspectival prejudice," which will help us understand why Augustine thinks animal death *appears* bad.

"TEMPORAL BEAUTY"

As we will explore further in the next chapter, Augustine holds that *human* death came into existence as a consequence of the fall of Adam and Eve. But he does not regard this event as the origin of death as such. Thus, when Augustine describes the effects of the human fall, he does not envision that Adam and Eve through their act of sin spread death and corruption to the animal kingdom, spoiling an unspotted, deathless, herbivorous environment. Just the opposite: at their fall, Adam and Eve "contract" the death that was already present around them in the animal kingdom. Referring to their eating the fruit in Genesis 3, Augustine writes:

> By this deed, in fact, they forfeited the wonderful condition, which was to be bestowed upon them through the mystical virtue in the tree of life. In this condition it would have been impossible for them to be tried by disease or altered by age. . . . When they forfeited this condition, then, their bodies contracted that liability to disease and death which is present in the flesh of animals.[9]

But if human sin did not introduce death and disease to the world, why are they a part of God's good creation?

---

[8]Obviously, other considerations will be needed to develop a full response to the challenge of natural evil. I trace out some of the literature in this vein, and provide a fourfold taxonomy of different responses to the challenge of prehuman natural evil, in "On the Fall of Angels and the Fallenness of Nature: An Evangelical Hypothesis Regarding Natural Evil," *Evangelical Quarterly* 87.2 (2015): 114-36.

[9]*De Genesi ad litteram* 11.32.42 (CSEL 28:1, 365-66).

Augustine is well aware of this problem. In *The City of God* he references how the Manichaeans find fault with "fire, frost, and wild beasts" because they threaten "this thin-blooded and frail mortality of our flesh."[10] Augustine faults the Manichaeans for failing to consider how admirable these things are in their own natures, and how "beautifully adjusted" to the rest of creation.[11] Moreover, their threat to humanity often depends on their misuse—even many poisons, Augustine points out, are "wholesome and medicinal" when used properly.[12] In addition, he faults the Manichaeans for breaching humility by passing judgment on what they do not understand: "Divine providence admonishes us not foolishly to vituperate things, but to investigate their utility with care."[13] When we cannot find any utility in a part of God's creation, we should assume that there is one that lies beyond our knowledge. In fact, he asserts that the very concealment of the use of something may be for the purpose of "a levelling of our pride."[14]

But as he proceeds, Augustine goes beyond merely cautioning against judgments of the animal kingdom. He is not reluctantly at peace with predatory animals, nor does he regard insect life as a neutral feature of God's world. Rather, he affirms them as a beautiful part of the way God has made the world, for which God should be praised. In particular, he develops a vision of the world in which the passing into nonexistence of some created things, including animals, displays the beauty of successive seasons in God's creation. For instance, he maintains:

> It is ridiculous to condemn the faults of beasts and of trees, and other such mortal and mutable things as are void of intelligence, sensation, or life, even though these faults should destroy their corruptible nature; for these creatures received, at their Creator's will, an existence fitting them, by passing away and giving place to others, to secure that

[10] *De civitate Dei* 11.22 (CSEL 40:1, 542-43).
[11] *De civitate Dei* 11.22 (CSEL 40:1, 543).
[12] *De civitate Dei* 11.22 (CSEL 40:1, 543).
[13] *De civitate Dei* 11.22 (CSEL 40:1, 543).
[14] *De civitate Dei* 11.22 (CSEL 40:1, 543).

> lowest form of beauty, the beauty of seasons, which in its own place is a requisite part of the world.[15]

What does he mean by the "lowest form of beauty" here? He is drawing on the immutable/mutable distinction surveyed in chapter 1—the spiritual creation shares in God's immutability, while the lower creation is mutable and thus exists by the passing of one thing into another. Thus, when commenting on God's instruction of the animals to "multiply" in Genesis 1:22 in his literal commentary, Augustine explains that "because they were created weak and mortal, they might preserve their kind by giving birth."[16] Then, in the context of commenting on the assertion "and God saw that it was good" throughout Genesis 1, Augustine articulates a distinction between abiding and transient created objects: "Some things, you see, abide by soaring over the whole rolling wheel of time in the widest range of holiness under God; while other things do so according to the limits of their time, and thus it is through things giving way to and taking the place of one another that the beautiful tapestry of the ages is woven."[17] Augustine sees this process of mutable creatures passing away and replacing one another as having an inferior kind of beauty—but is beautiful nonetheless.

In his *On Free Choice of the Will*, Augustine develops this view:

> It is foolish for us to say that temporal objects ought not to pass away. They have been placed in the order of the universe in such a way that, unless they do pass away, future objects cannot succeed to past ones, and only thus can the whole beauty of times past, present, and future be accomplished in their own kind. They use what they have received and return it to Him to whom they owe their existence and greatness. The man who grieves that these things pass away ought to listen to his own speech, to see if he thinks the complaint that he makes is just and proceeds from prudence. . . . In the case of objects which pass away because

[15]*De civitate Dei* 12.4 (CSEL 40.1, 571).
[16]*De Genesi ad litteram liber unus imperfectus* 15.50 (CSEL 28:1, 494).
[17]*De Genesi ad litteram* 1.8.14 (CSEL 28:1, 11).

> they have been granted only a limited existence in order that everything may be accomplished in its time, no one is right to blame this deficiency.[18]

Here Augustine reasons that God has the right to withdraw existence from temporal objects, since he was the one who gave it in the first place. He also observes that temporal objects must pass away in order to make space for new ones; for instance, if no animal species ever went extinct, far fewer kinds of animals would be created, and if no individual animals ever died, far fewer actual animals would be created. All this entails, for Augustine, that we are wrong to sit in judgment on the passing away of created objects as though this is a blameworthy aspect of God's creation. On the contrary, this seasonal progression in God's creation contains a kind of beauty ("the beautiful tapestry of the ages," as he put it).

This language of seasonal beauty suggests as a possible analogy to Augustine's idea the four seasons of the year: winter, spring, summer, fall. There is a kind of beauty in each of these seasons, considered in isolation: white snow is beautiful, for instance, as are colorful autumn leaves. But there is also a different and deeper kind of beauty in the larger whole of which they are all a part, and each particular season would be less beautiful if it were extracted from this larger pattern. For instance, we appreciate autumn leaves precisely because they follow and replace green summer leaves. But this larger seasonal beauty can only be accomplished by the changing of one season into another—that is, by the cessation of the lesser beauty of each part. Unless the leaves fall off, they cannot grow back again; unless the snow melts, we have no spring greenery.

Augustine applies this language of transient/temporal/seasonal beauty not only to the animal kingdom wholesale, but to various particular creatures, including the ones we often find unpleasant. At one point in his *On True Religion* he refers to smaller creatures such as

[18]*De libero arbitrio* 3.15 (CSEL 74, 125-26).

insects as "breakable beauties" (*pulchritudines corruptibiliores*), claiming, "Let us not be surprised at my still calling them beauties," citing Romans 13:1 ("all order is from God") to establish that beauty is inherent in all order.[19] Augustine's admission that worms are inferior to human beings does not hinder his ability to see their good qualities: "We are bound to admit that a weeping man is better than a rejoicing worm, and yet I am not lying when I say that I can say volumes in praise of the worm."[20] Specifically, Augustine praises the worm's "bright color, the smooth round shape of its body, the first sections fitting into the middle ones, the middle ones fitting into those at the end, all observing the aim of unity according to the lowliness of their nature, nothing being formed from one part which does not correspond in parallel dimensions with another."[21] Augustine also marvels at its ability to move rhythmically, aim toward what suits it, and avoid danger. Perhaps anticipating that not all of his readers will share his intuition that worms are beautiful, Augustine then goes even further, asserting that "many have spoken most truly and eloquently in praise of ashes and dung."[22]

In his sermons as well, Augustine praises the orderly nature of "the tiniest and least of living mites . . . the seeds, roots, trunks, branches, leaves, flowers, fruits of countless trees and herbs, from nature's secret stores."[23] He sees all of these parts of God's creation as evidences of the magnificent and "omnificent"[24] wisdom of God. He proceeds to warn his listeners:

> It is surely the last word in absurdity to deny in great matters that divine provision and forethought which we admire in small ones. . . .

---

[19]*De vera religione* 41.77 (CSEL 77, 56).
[20]*De vera religione* 41.77 (CSEL 77, 56).
[21]*De vera religione* 41.77 (CSEL 77, 56).
[22]*De vera religione* 41.77 (CSEL 77, 56).
[23]Sermon *De providentia Dei*, in *Sermons Discovered Since 1990*, ed. John E. Rotelle, trans. Edmund Hill, The Works of Saint Augustine: A Translation for the 21st Century (Hyde Park, NY: New City Press, 1997), 58.
[24]Latin: *omnifica*; he apparently develops this word as a superlative of "magnificent."

> Let us therefore please have no hesitation in believing that what seems to be messy and disordered in human affairs is governed, not by no plan at all, but rather by an altogether loftier one, and by a more all-embracing divine order that can be grasped by our human littleness.[25]

Evidently it is not just post-Darwinian Christians who can be disturbed by the apparent *messiness* and *disorder* of the natural world—and the impression of purposelessness it engenders. The differences between Augustine's time and ours therefore need not render less relevant his warning that God may have "loftier" and "more all-embracing" purposes that lie beyond the comprehension of "human littleness." On this point, Augustine represents a different set of instincts than those modern skeptics who wield the quip of J. B. S. Haldane to undermine the plausibility of theism: "God must have an inordinate fondness of beetles."[26]

In his finished literal commentary on Genesis, Augustine poses the question of whether "the minutest animals" were created in the original establishment of things, or whether they are the result of the "putrefaction consequent upon material things being perishable."[27] He notes that such animals are typically generated from the sores of living bodies, or from garbage or corpses or some other rotting material, and affirms that "we cannot possibly say that there are any of them of which God is not the creator."[28] He then establishes this claim by arguing that all creatures, however small, reflect the wisdom of the Creator. Far from being outside the scope of God's goodness and wisdom, the activity of insects gives us all the more reason to praise him:

> All things, after all, have in them a certain worth or grace of nature, each of its own kind, so that in these minute creatures there is even

[25]Sermon *De providentia Dei*, in *Sermons Discovered Since 1990*, 58.

[26]The quotation is disputed; see the discussion in K. N. Ganeshaiah, "Haldane's God and the Honoured Beetles: The Cost of a Quip," *Current Science* 74.8 (1998): 656-60.

[27]*De Genesi ad litteram* 3.14.22 (CSEL 28:1, 79).

[28]*De Genesi ad litteram* 3.14.22 (CSEL 28:1, 79).

> more for us to wonder at as we observe them, and so to praise the almighty craftsman for them more rapturously than ever. . . . If we pay close attention we are more amazed at the agile flight of a fly than at the stamina of a sturdy mule on the march; and the cooperative labors of tiny ants strike us as far more wonderful than the colossal loads that can be carried by camels.[29]

Augustine proceeds to argue that some insects were created at the original creation, while those that are generated from the bodies of animals were not, unless they were somehow "seeded" into the animals to arise later.[30] He does not develop this ambiguous suggestion or pose an alternative possibility.

Augustine's instinct to praise God for insects is also evident in the *Confessions*, where he bemoans the fact that worship is not his first instinct while he is sitting at home watching flies getting eaten by lizards and spiders:

> What excuse can I make for myself when often, as I sit at home, I cannot turn my eyes from the sight of a lizard catching flies or a spider entangling them as they fly into her web? Does it make any difference that these are only small animals? It is true that the sight of them inspires me to praise you for the wonders of your creation and the order in which you have disposed all things, but I am not intent upon your praises when I first begin to watch.[31]

Elsewhere, Augustine extends his defense of the goodness of insects to larger predatory animals, and applies his conception of "temporal beauty" not only to animal death but to animal pain and predation as well. Thus, Augustine maintains that animal predation should not be regarded as bad since it plays a larger role in God's good and wise government of the world. For instance, in his literal commentary on Genesis, he tackles the "commonly asked" question of

---

[29]*De Genesi ad litteram* 3.14.22 (CSEL 28:1, 79-80).
[30]*De Genesi ad litteram* 3.14.23 (CSEL 28:1, 80).
[31]*Confessiones* 10.35 (CSEL 33, 269-70).

whether poisonous and dangerous animals were created after human sin as a punishment or had been created harmless and began to do sinners harm after the fall.[32] He sees the peaceful coexistence of unfallen humans and predatory animals as entirely possible, citing the examples of Daniel in the lions' den in Daniel 6 and Paul being unharmed by the deadly viper in Acts 28.[33] Later, he envisions the possibility of Adam in his pre-fallen state guarding Eden against wild beasts—he finds it implausible that in this scenario the beasts could be a threat to him, but does not rule it out.[34] But then Augustine anticipates this question: "Someone is going to say: 'Then why do beasts injure one another, though they neither have any sins, so that this kind of things could be called punishment, nor by such trials do they gain at all in virtue?'"[35]

Here Augustine has stated the problem of "natural evil" as applied to the animal kingdom in strikingly similar terms to those of the contemporary philosophical discussion.[36] His response to this question involves an appeal to the principle of "temporal beauty" explored above:

> For the simple reason, of course, that some are the proper diet of others. Nor can we have any right to say, "There shouldn't be some on which others feed." All things, you see, as long as they continue to be, have their own proper measures, numbers and destinies. So all things, properly considered, are worthy of acclaim; nor is it without some contribution in its own way to the temporal beauty of the world that they undergo change by passing from one thing into another. This may

---

[32]*De Genesi ad litteram* 3.15.24 (CSEL 28:1, 80-81).

[33]*De Genesi ad litteram* 3.15.24 (CSEL 28:1, 81).

[34]*De Genesi ad litteram* 8.10.21 (CSEL 28:1, 246).

[35]*De Genesi ad litteram* 3.16.25 (CSEL 28:1, 81).

[36]E.g., C. S. Lewis in *The Problem of Pain*: "The problem of animal suffering is appalling; not because the animals are so numerous (for, as we have seen, no more pain is felt when a million suffer than when one suffers) but because the Christian explanation of human pain cannot be extended to animal pain. So far as we know beasts are incapable either of sin or virtue: therefore they can neither deserve pain nor be improved by it" (*The Complete C. S. Lewis Signature Classics* [New York: HarperOne, 2002], 628).

> escape fools; those making progress have some glimmering of it; to the perfect it is as clear as daylight.[37]

The forcefulness of Augustine's conviction here is striking. He not only defends animal predation, but he rebukes the tendency to sit in judgment on it as an illegitimate and even foolish impulse.

Ultimately, Augustine divides the entire animal kingdom into three categories in their relation to humanity: some are beneficial, some are pernicious, and some are superfluous. He argues that pernicious animals serve a spiritual purpose in relation to humanity, helping us to long for heaven. They are here either to punish or frighten us, "so that instead of loving and desiring this life, subject to so many dangers and trials, we should set our hearts on another better one, where there is total freedom from all worries and anxieties."[38] As for the superfluous ones, we should not call them into question, for "they contribute to the completion of this universe, which is not only much bigger than our homes, but much better as well; God manages it after all, much better than any of us can manage our homes."[39] The introduction of this comparison between the universe and a human home, though serving the rhetorical effect of cautioning human judgments against creation, also reflects Augustine's vision of the universe as a place of order and intrinsic value, which God "manages" wisely. Augustine concludes by emphasizing that although different creatures call for a different response from human beings, they should all cause us to praise the Creator:

> So then, make use of the useful ones, be careful with the pernicious ones, let the superfluous ones be. In all of them, though, when you observe their measure and numbers and order, look for the craftsman. . . . In this way you will perhaps find more genuine satisfaction when you praise God in the tiny little ant down on the ground, than when you are crossing a river high up, let us say, on an elephant.[40]

---

[37]*De Genesi ad litteram* 3.16.25 (CSEL 28:1, 81-82).

[38]*De Genesi contra Manichaeos* 1.16.26 (CSEL 91, 93).

[39]*De Genesi contra Manichaeos* 1.16.26 (CSEL 91, 93).

[40]*De Genesi contra Manichaeos* 1.16.26 (CSEL 91, 93-94).

Augustine admits his own personal uncertainty of why God made certain animals. But he also maintains that this does not inhibit his ability to find them beautiful: "I, however, must confess that I have not the slightest idea why mice and frogs were created, and flies and worms; yet I can still see that they are all beautiful in their own specific kind."[41] He emphasizes that there is a beauty in the organization of all living creatures, even the tiniest ones: "There is not a single living creature, after all, in whose body I will not find, when I reflect upon it, that its measures and numbers and order are geared to a harmonious unity."[42] As with the quotation from the literal commentary 3.16.25 above, the language of *measure and number and order* is drawn from Wisdom 11:20, though here Augustine has substituted "order" for "weight," in line with his emphasis. If the Manichaeans were to observe this orderliness in even the tiniest creatures, Augustine argues, "they wouldn't go on boring us to death, but by reflecting themselves on all such beauties from the highest to the lowest would in all cases praise God the craftsman."[43] Augustine acknowledges that the lower animals cause a certain kind of offense to humanity, but he introduces a distinction between that which is offensive to reason and that which is merely offensive to our "carnal senses."[44] The lower animals, such as mice and frogs, or insects, are only offensive to our carnal senses, not to reason.

This distinction relates to a common theme in Augustine's treatment of animal death: the danger of making self-referential judgment. In *The City of God*, he warns against reducing the value of animals to their utility for humanity, and in particular denigrating those creatures we would foolishly want to abolish from existence altogether: "Who, e.g., would not rather have bread in his house than mice, gold

[41]*De Genesi contra Manichaeos* 1.16.26 (CSEL 91, 92).
[42]*De Genesi contra Manichaeos* 1.16.26 (CSEL 91, 92).
[43]*De Genesi contra Manichaeos* 1.16.26 (CSEL 91, 93).
[44]*De Genesi contra Manichaeos* 1.16.26 (CSEL 91, 93).

than fleas?"[45] For Augustine, finding fault with such creatures is not a judgment resulting from wisdom, but rather a judgment resulting from pleasure, need, or desire. Thus, he maintains, if we desire to get rid of those animals that annoy us (such as mice or fleas), we may be motivated impulsively, failing to take into account their role in the larger created order.[46]

Augustine also emphasizes that animal death provides humanity with many "salutary admonitions" that may contribute to our understanding of salvation. Specifically, he argues that by observing the struggle of animals to secure their bodily, temporal welfare, we might learn what trouble we should take in pursuing our everlasting spiritual welfare. We can learn such lessons from the entire animal kingdom, "from the biggest elephants down to the smallest little worms."[47] He emphasizes in particular the salutary value of bodily pain in animals, which he regards as a "great and wonderful power of the soul."[48]

Elsewhere, Augustine pushes back further against the Manichaean intuition that pain is necessarily evil. In his *The Nature of the Good*, he argues that pain exists precisely because our natures are good: "Even pain, however, which some persons consider especially evil, whether it is in the mind or in the body, cannot exist except in good natures."[49] Augustine observes that pain results from resisting that which is destructive—when the will resists a greater power, it produces mental pain, and when the senses resist a more powerful body, it produces bodily pain. Augustine therefore distinguishes between different results of pain: "When it is forced to something better, pain is beneficial; when it is forced to something worse, it is harmful."[50]

---

[45]*De civitate Dei* 11.16 (CSEL 40.1, 535).
[46]*De civitate Dei* 11.16 (CSEL 40.1, 535).
[47]*De Genesi ad litteram* 3.16.25 (CSEL 28:1, 82).
[48]*De Genesi ad litteram* 3.16.25 (CSEL 28:1, 82).
[49]*De natura boni* 20 (CSEL 25.2, 863).
[50]*De natura boni* 20 (CSEL 25.2, 863).

Nonetheless, Augustine emphasizes that there are many worse evils that occur without pain.[51] In his allegorical commentary on Genesis, similarly, Augustine speculates that pain in childbearing is not itself a consequence of sin but rather a function of being mortal, since female animals also give birth in pain.[52]

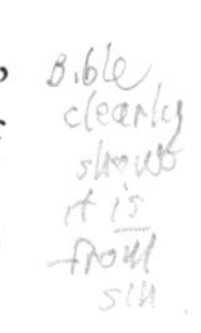

When Augustine turns in his literal commentary to address whether thorns, thistles, and trees that produce no fruit existed before the fall, he again urges caution. Quoting the reference to "fruit-bearing trees" in Genesis 1:11, Augustine stipulates that the word "fruit" refers to the things' use for those who enjoy, and emphasizes that there are many different kinds of uses that the things of the earth have. He distinguishes between obvious uses that people can observe for themselves and the hidden ones we must learn from the experts. We must therefore be careful in our judgment of what a "fruitless" tree is.[53] When it comes to thorns and thistles, Augustine quotes the curse of Genesis 3:18: "thorns and thistles it shall bring forth for you."[54] But Augustine cautions that this verse does not give anyone warrant to conclude that thorns and thistles began with the fall. He had articulated this view in his earlier allegorical work in response to Manichaean objections.[55] Here he reasons, "Many uses can be found for seeds of this kind too, and so they could have had their place without any penal inconvenience to the man."[56] Augustine sees several ways to understand this possibility. For instance, thorns and thistles could have been produced specifically in the field that man was given to cultivate. Or, they could have been produced in all places but simply become noxious *to the man* after the fall.[57] In all this, Augustine displays caution about exaggerating the effects of the fall as producing

[51]*De natura boni* 20 (CSEL 25.2, 863).
[52]*De Genesi contra Manichaeos* 2.19.29 (CSEL 91, 150).
[53]*De Genesi ad litteram* 3.18.27 (CSEL 28:1, 83).
[54]*De Genesi ad litteram* 3.18.28 (CSEL 28:1, 83).
[55]*De Genesi contra Manichaeos* 1.13.19 (CSEL 91, 85-86).
[56]*De Genesi ad litteram* 3.18.28 (CSEL 28:1, 83-84).
[57]*De Genesi ad litteram* 3.18.28 (CSEL 28:1, 84).

an absolute contrast—he sees the fall, not as originating human pain and labor, but as aggravating them.

Augustine's emphasis here is different from the approach of other fathers, such as Basil and Ambrose, who interpreted the threat of "thorns and thistles" in Genesis 3:18 as a metaphor for the deterioration of postlapsarian life.[58] Yet Augustine's view became more common in his wake of influence—it is picked up, for instance, in the fifth century by Claudius Marius Victorius and the Roman Emperor Avitus, and persisted into the medieval era (e.g., in Lombard's *Sentences*) despite Augustine's earlier opposing view (from his *De Genesi contra Manichaeos*) being mistakenly attributed to his literal commentary in Bede's writings.[59] His broader conviction that animal death was not introduced to the world by the fall was also widely shared. In the medieval period, Thomas Aquinas quoted Augustine in order to oppose the view that there ought to have been nothing injurious to humanity before the fall.[60] For Aquinas, "The nature of animals was not changed by man's sin, as if those whose nature now it is to devour the flesh of others, would then have lived on herbs, like the lion and falcon."[61] Thus, Augustine is not alone in the ways that his intuitions about animal death differ from some modern views; he represents a broader tradition of thought.

"PERSPECTIVAL PREJUDICE"

But all this raises a question: If animal death contributes to the beauty of the world, why does it trouble us so much? If it is beautiful, why do

[58]See Karla Pollmann, "'And Without Thorn the Rose'? Augustine's Interpretation of Genesis 3:18 and the Intellectual Tradition," in *Genesis and Christian Theology*, ed. Nathan MacDonald, Mark W. Elliott, and Grant Macaskill (Grand Rapids: Eerdmans, 2012), 217-18.

[59]Pollmann, "'And Without Thorn the Rose'?," 222-25. In Bede's commentary on Genesis 3:17-18, a quotation from Augustine's *De Genesi contra Manichaeos* is apparently falsely attributed to Augustine's *De Genesi ad litteram*.

[60]*Summa Theologica* I, q. 72, a. 1, ad. 6, trans. Fathers of the English Dominican Province (Notre Dame, IN: Christian Classics, 1948), 352.

[61]Thomas Aquinas, *Summa Theologica* I, q. 96, a. 1, ad 2, 486.

we often find it ugly? To answer this, we must consider a broader principle of Augustine's theodicy, which we will call, following William Mann, "perspectival prejudice." Mann does not spend much time developing this idea but defines it as "failing to see how local privations, especially the ones that affect us, contribute to the good of the whole."[62] The challenge of perspectival prejudice is a huge part of Augustine's thought on this topic, and comes up again and again in his writings.

That this idea is present in Augustine's mind following his rejection of Manichaeism is evidenced by his claim in *On Order*, the first book written after his conversion:

> Whoever narrow-mindedly considers this life by itself alone is repelled by its enormous foulness, and turns away in sheer disgust. But, if he raises the eyes of the mind and broadens his field of vision and surveys all things as a whole, then he will find nothing unarranged, unclassed, or unassigned to its own place.[63]

Here Augustine emphasizes the *organization* of creation as an answer to its unpleasantness—what undercuts the reaction of "disgust" at the "foulness" of the world is specifically the recognition that there is nothing that is "unarranged, unclassed, or unassigned to its own place." The acuteness of Augustine's sensitivity to this challenge is reflected in his language here ("enormous foulness . . . sheer disgust")—whatever else we might say, we cannot dismiss Augustine as failing to take seriously the weight of the problem.

As he develops this notion throughout his subsequent writings, Augustine draws together his rebuke against narrowness of vision with a rebuke against self-referential judgment. In other words, the

[62]William E. Mann, "Augustine on Evil and Original Sin," in *The Cambridge Companion to Augustine*, ed. David Vincent Meconi and Eleonore Stump, 2nd ed. (Cambridge: Cambridge University Press, 2014), 103.

[63]*De ordine* 2.4.11 (CSEL 63, 154); translation drawn from N. Joseph Torchia, *Creatio Ex Nihilo and the Theology of Augustine: The Anti-Manichaean Polemic and Beyond*, American University Studies 205 (New York: Peter Lang, 1999), 172.

problem of perspectival prejudice involves not merely people failing to interpret local phenomena in relation to the whole, but people failing to take into account their own location and involvement in what they are judging. Thus, after describing animal and plant death as constituting the "lowest form of beauty" in *The City of God*, Augustine observes that "of this order the beauty does not strike us, because by our mortal frailty we are so involved in a part of it, that we cannot perceive the whole, in which these fragments that offend us are harmonized with the most accurate fitness and beauty."[64] Here the problem of not seeing the whole is drawn together with the problem of our own self-involvement in the situation we are seeking to judge: it is because our own "mortal frailty" involves us in this form of beauty that we cannot see the bigger picture. Augustine proceeds by rebuking the rashness of the Manichaeans, who judge animals by their utility rather than by their nature:

> But in this way of estimating, they may find fault with the sun itself; for certain criminals or debtors are sentenced by the judges to be set in the sun. Therefore, it is not with respect to our convenience or comfort, but with respect to their own nature, that the creatures are glorifying to their Artificer. . . . We must not listen, then, to those who praise the light of fire but find fault with its heat, judging it not by its nature, but by their comfort or convenience. For they wish to see, but not to be burnt.[65]

Augustine proceeds to argue that God is to be praised for every nature he has made rather than blamed for their faults.[66]

In his allegorical commentary on Genesis, Augustine develops a metaphor to emphasize this same point. In context, he is dealing with the Manichaean objection that God has made so many animals that are not necessary for human beings, and that many of

---

[64]*De civitate Dei* 12.4 (CSEL 40.1, 571).
[65]*De civitate Dei* 12.4 (CSEL 40.1, 572).
[66]*De civitate Dei* 12.5 (CSEL 40.1, 572-73).

them are "pernicious and to be feared."[67] His response is worth quoting at length:

> But when they say things like that, they are failing to understand how all these things are beautiful to their maker and craftsman, who has a use for them all in his management of the whole universe which is under the control of his sovereign law. After all, if a layman enters a mechanic's workshop, he will see many instruments there whose purpose he is ignorant of, and which, if he is more than usually silly, he thinks are superfluous. What's more, if he carelessly tumbles into the furnace, or cuts himself on a sharp steel implement when he handles it wrongly, then he reckons that there are many pernicious and harmful things there too. The mechanic, however, who knows the use of everything there, has a good laugh at his silliness, takes no notice of his inept remarks, and just presses on with the work in hand. And yet people are so astonishingly foolish, that with a human craftsman they won't dream of objecting to things whose function they are ignorant of, but will assume when they notice them that they are necessary, and put there for various uses, and yet will have the audacity, looking round this world of which God is the acknowledged founder and administrator, to find fault with many things which they cannot see the point of.[68]

Here Augustine goes beyond merely rebuking self-referential judgments about things like predatory animals, inviting as an alternative the consideration of *God's* judgments; some parts of the natural order might seem superfluous or even dangerous to us, and yet appear "beautiful to their maker and craftsman." God is the mechanic, we are the layman, and the world is God's workshop. Augustine emphasizes here the limitation of human knowledge and the qualitative difference between God's relation to the natural world and our relation to it. To put it simply, God knows how the world works and we do not. Sitting

---

[67] *De Genesi contra Manichaeos* 1.16.25 (CSEL 91, 91).

[68] *De Genesi contra Manichaeos* 1.16.25 (CSEL 91, 91-92).

in judgment on the world is therefore a careless error resulting from audacity and silliness, and inviting laughter as "astonishingly foolish."

Throughout his writings, Augustine uses a number of other metaphors to develop his conception of perspectival prejudice. In his allegorical commentary on Genesis, for example, he observes that individual created works are called "good," while the entire creation is called "very good."[69] From this he develops the principle that the beauty of the whole world is greater than that of the individual parts, comparing this to a human body, in which individual parts (like eyes, nose, cheeks, etc.) are beautiful, but the beauty of the entire body working together is greater. In the created world, "Such is the force and power of completeness and unity, that many things, all good in themselves, are only found satisfying when they come together and fit into one universal whole. The universal, the universe, of course, takes its name from unity."[70] Augustine then introduces the further metaphor of a speech, in which the individual parts are meaningless unless they are heard in relation to the whole. This inability to interpret the parts in relation to the whole is at the root of the Manichaean error: "If the Manichaeans would only consider this truth, they would praise God the author and founder of the whole universe, and they would fit any particular part that distresses them because of our mortal condition into the beauty of the universal whole, and thus would see how God made all things not only good, but very good."[71]

This metaphor recurs in *The City of God*, where Augustine compares evil in creation to verbal antitheses (which he also calls oppositions or contrapositions); these he regards as "among the most elegant of the ornaments of speech."[72] After quoting from 2 Corinthians 6:7-10 as an example of this literary phenomenon, he concludes: "As,

[69]*De Genesi contra Manichaeos* 1.21.32 (CSEL 91, 100).
[70]*De Genesi contra Manichaeos* 1.21.32 (CSEL 91, 100).
[71]*De Genesi contra Manichaeos* 1.21.32 (CSEL 91, 100-101).
[72]*De civitate Dei* 11.18 (CSEL 40.1, 537).

then, these oppositions of contraries lend beauty to the language, so the beauty of the course of this world is achieved by the opposition of contraries, arranged, as it were, by an eloquence not of words, but of things."[73]

In his *On Music*, written around 391, Augustine combines the metaphors of a statue, a soldier, and a poem to convey the temporal beauty and "harmonious succession" of God's creation:

> Terrestrial things are subject to celestial, and their time circuits join together in harmonious succession for a poem of the universe.
>
> And so many of these things seem to us disordered and perturbed, because we have been sewn into their order according to our merits, not knowing what a beautiful thing divine providence purposes for us. For, if someone should be put as a statue in an angle of the most spacious and beautiful building, he could not perceive the beauty of the building he himself is a part of. Nor can the soldier in the front line of battle get the order of the whole army. And in a poem, if syllables should live and perceive only so long as they sound, the harmony and beauty of the connected work would in no way please them. For they could not see or approve the whole, since it would be fashioned and perfected by the very passing away of these singulars.[74]

These images once again signal the danger of self-referential judgment: it is the *location* of the statue in an angle of the building and of the soldier in the front line of battle that makes a detached, comprehensive view impossible. Similarly, we are a part of the universe ("sewn" into it), and thus do not see it as God (who is not a part of it) sees it. This is why things seem to us "disordered and perturbed"—it would not appear this way to us if we did not suffer from this limited perspective, but instead could see "knowing what a beautiful thing divine providence purposes for us."

---

[73]*De civitate Dei* 11.18 (CSEL 40.1, 538).

[74]Augustine, *On Music* 6.11, ed. Hermigild Dressler, trans. Robert Catesby Taliaferro, vol. 4, The Fathers of the Church (Washington, DC: Catholic University of America Press, 1947), 355.

If the statue and building metaphors here draw attention to our spatial limitations, the poetry metaphor draws attention to our temporal limitations. This metaphor recurs in Augustine's *On True Religion*, which was written around the same time as *On Music*. Here Augustine contrasts the resurrection body, which will not be subject to decay, with our current bodily condition, which belongs to "the least and lowest beauty of bodies."[75] This is a form of beauty that is "carried along in a successive order . . . because it cannot have everything at once and all together; but, while some things give way and others take their place, they fill up the number of time-bound forms and shapes into a single beauty."[76] Then Augustine uses an illustration to argue that this lower form of beauty, although much inferior to our resurrection status, is not on that account bad:

> The fact that all this is transitory does not make it evil. For this is the way in which a line of poetry is beautiful, even though two syllables of it cannot possibly in any way be spoken simultaneously. I mean that the second one cannot be pronounced unless the first one has passed away, and so in due course you reach the end, so that when the last syllable is heard, without the previous ones being heard simultaneously, it still completes the form and beauty of the meter by being woven in with the previous ones.[77]

What Augustine particularly emphasizes in this poetry metaphor is the *successive* nature of God's purposes and activity in his creation. A few chapters later, he will refer to "a certain time-bound method of healing" of the human race, and how God gradually brings about his purposes for humanity over the course of many different ages.[78] Thus, Augustine conceives of God's purposes for the natural order as a kind of poem, proceeding verse by verse toward its end. To object to particular

[75]*De vera religione* 21.41 (CSEL 77, 29).
[76]*De vera religione* 21.41 (CSEL 77, 29).
[77]*De vera religione* 22.42 (CSEL 77, 29-30).
[78]*De vera religione* 24.45 (CSEL 77, 32); cf. 26.48-49 (CSEL 77, 34-35).

instances of pain or discord in the world as evil is therefore comparable to criticizing the literary quality of a poem before you've heard how it ends. Augustine even goes so far as to assert that those who have no appreciation for this aspect of divine providence are "behaving as absurdly as if someone in the recitation of some well-known poem wanted to listen to just one single syllable all the time."[79]

Augustine does not share the intuition occasionally found in contemporary views that if creation is good, it must be perfect. Nor does he understand the goodness of creation in static terms. Rather, there is a developmental component to his conception of the goodness of creation. This draws on his twofold conception of the goal of creation, as we explored in chapter 1. At one point in his finished Genesis commentary, for instance, in connection to his conception of *rationes seminales*, he asks, "If God were to make anything imperfect, which he would then himself bring to perfection, what would be reprehensible about such an idea?"[80] The logic here involves the assumption that the ultimate end of God's purposes must be considered in order to make judgments about the process leading to that end. This is not exactly tantamount to the claim that the ends justify the means—for one thing, Augustine does not seem to think of the means as "bad" and in need of *justification* so much as "imperfect" and in need of *development*, and this is an important distinction. Nonetheless, his comment does seem to envision a developmental model of creation, and the need to consider its ultimate end to determine and understand its goodness.

Later in *On True Religion* he returns to spatial imagery. Again here Augustine casts various unpleasant phenomena in nature as a lower form of beauty—indeed, at the very edge of what may be called beautiful: "Divine providence is at hand to show us both that this lowest kind of beauty is not bad . . . and yet to make it clear that it is at the

[79]*De vera religione* 22.43 (CSEL 77, 30).
[80]*De Genesi ad litteram* 2.15.30 (CSEL 28:1, 210).

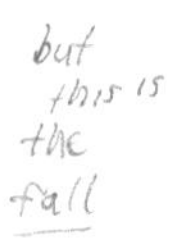

outermost edge of beauty by mixing in with it pains and diseases and distortion of limbs and dark coloring and rivalries and quarrels of spirits."[81] But God uses these unpleasant elements to provide warnings and training to the good, so that "in this way all are directed by their functions, their duties and their ends toward the beauty of the whole universe, so that if what shocks us in the part is considered together with the whole, it gives us entire satisfaction." Augustine then introduces new illustrations to advance this point:

> After all, in making a judgment on a building we ought not to consider just one single corner; or in assessing the beauty of human beings just their hair; or with good public speakers just the movements of their fingers; or with the course of the moon just its changes during three days. These things, you see, which are at the lowest level precisely because, while complete in themselves, as parts they are incomplete, are to be considered with the wholes they are part of, if we wish to make a right judgment.[82]

A few sentences later, he goes on to add the further image of how "the color black in a picture becomes beautiful within the whole," using this to describe God's providential oversight over good and evil. These metaphors emphasize the utter foolishness of passing judgment on God's creation, even those aspects that are at the "outermost edge of beauty." To judge the beauty of human beings only by looking at their hair, or to judge a speech only by looking at the speaker's fingers—these are not only imperfect judgments, but utterly absurd ones. So Augustine urges caution with our judgments when we are shocked by particular aspects of God's creation.

Augustine's conception of evil as essentially a privation plays an important role in his appeal to the principle of perspectival prejudice. In his unfinished literal commentary on Genesis, he refers to those

[81]*De vera religione* 40.75 (CSEL 77, 54).
[82]*De vera religione* 40.76 (CSEL 77, 55).

"lacks and absences" that "have their due place in the total pattern of things designed and controlled by God."[83] He uses metaphors to develop how such privations contribute to the good of God's creation, comparing them to silences in singing, which "contribute to the overall sweetness of the whole song," and shadows in pictures, which "mark out the more striking features, and satisfy by the rightness not of form but of order and arrangement."[84] He develops a distinction as a consequence of these images: "So then, there are some things that God both makes and controls or regulates, while there are some that he only regulates."[85] This distinction allows Augustine to attribute the organization of the entire universe to God without thereby making him the author of evil. As an example, he appeals not only to just human beings (whom God both makes and regulates) and unjust human beings (whom God only regulates), but also the broader realm of nature: "So it is that he both makes and regulates the forms and natures of different species, while as for the shortcomings of forms and the defects of natures, he does not make them but only regulates them."[86] He appeals to the existence of darkness as opposite light as another example, concluding that "thus there is beauty in every single thing, with him making it; and with him arranging them in regular order there is beauty in all things together."[87]

Another component of Augustine's principle of perspectival prejudice is his emphasis on the hierarchical organization of creatures. Not only does he not assume that a good creation must be a perfect one, he also does not assume that if God makes all things good, he must make them all *equally* good. Rather, Augustine maintains that it is perfectly appropriate for God to make some natures better than others. After all, he reasons, a world with varying kinds of goods is,

---

[83] *De Genesi ad litteram liber unus imperfectus* 5.25 (CSEL 28:1, 475).
[84] *De Genesi ad litteram liber unus imperfectus* 5.25 (CSEL 28:1, 475).
[85] *De Genesi ad litteram liber unus imperfectus* 5.25 (CSEL 28:1, 475).
[86] *De Genesi ad litteram liber unus imperfectus* 5.25 (CSEL 28:1, 475).
[87] *De Genesi ad litteram liber unus imperfectus* 5.25 (CSEL 28:1, 476).

on the whole, better than one in which everything is equally good. Thus, in the *Confessions*, Augustine contrasts the lower beauty of earth with the higher beauty of the spiritual creation, insisting that while beauty has been added to all of God's creation, it has not been distributed uniformly throughout.[88] In fact, these different kinds of beauty are so different from each other that one (heaven) is "close to yourself (God)," while the other (the earth) is "little more than nothing."[89] Augustine ridicules the insistence that God should have made all things equally good as an instance of perspectival prejudice, comparing this claim to those who reason, "Since the sense of sight is better than that of hearing, it would have been better to have four eyes and no ears."[90]

At the conclusion of his allegorical commentary on Genesis, Augustine draws together his principles of evil as privation, the hierarchy of creatures, and the superiority of the whole over the parts:

> *We* say that there is no natural evil, but that all natures are good, and God himself is the supreme nature and all other natures come from him; and all are good insofar as they exist, since God made all things *very good* (Gn 1:31), but ranged them in an order of graded distinctions, so that one is better than another; and in this way the whole universe is completed out of every kind of good thing, and with some of them being perfect, others imperfect, is itself a perfect whole.[91]

Again, for Augustine, that some objects in the world are perfect and others are imperfect is *itself* part of what constitutes the perfection of the whole. Similarly, as we hinted at earlier, in the *Confessions* he emphasizes this point by drawing on the distinction between the "good" of each thing and the "very good" of the entire creation in Genesis 1,[92] speaking of how God's individual works have both

---

[88] *Confessiones* 11.2 (CSEL 33, 311).
[89] *Confessiones* 11.7 (CSEL 33, 314).
[90] *De Genesi ad litteram* 11.8.10 (CSEL 28:1, 341).
[91] *De Genesis contra Manichaeos* 2.29.43 (CSEL 91, 171).
[92] *Confessiones* 13.32 (CSEL 33, 385).

"progress and decline, beauty and defect," and they all serve to contribute more greatly to the beauty of the whole.[93]

But Augustine goes even further than this. Not only are some creatures not as good as others, but some creatures have defects in themselves but are still good in relation to the entire creation. In his literal commentary, Augustine asserts that sin corrupts, but nonetheless those things corrupted by sin are still good in relation to the rest of creation: "Those things, however, which lose their comeliness by sinning, do not in the least for all that bring it about that they too are not good when rightly coordinated with the whole, with the universe."[94] Augustine even goes so far as to correlate God's ability to create good things with his ability to providentially rule over bad things: "God, after all, while being the best creator of natural things, is also the most just co-ordinator of sinners; so that even if things individually become deformed by transgressing, nonetheless the totality together with them in it remains beautiful."[95] Similarly, in *The City of God* he writes, "As the beauty of a picture is increased by well-managed shadows, so, to the eye that has skill to discern it, the universe is beautified by sinners, though, considered by themselves, their deformity is a sad blemish."[96] In his *Enchiridion*, Augustine claims, "In this universe even that which is called evil, well ordered and kept in its place, sets the good in higher relief, so that good things are more pleasing and praiseworthy than evil ones."[97] Elsewhere he makes this point with (yet another) metaphor: "No loser enjoys the wrestling matches in the games, and yet his defeat contributes to their success. And this, after all, is a kind of copy of the Truth. . . . In the same kind of way it is only godless and condemned souls who take no pleasure in the state and

[93]*Confessiones* 13.33 (CSEL 33, 385).
[94]*De Genesi ad litteram* 3.24.37 (CSEL 28:1, 92).
[95]*De Genesi ad litteram* 3.24.37 (CSEL 28:1, 92).
[96]*De civitate Dei* 11:23 (CSEL 40:1, 545).
[97]*Enchiridion* 3.11, in *On Christian Belief*, 278.

organization of this universe."[98] It is striking that Augustine not only can claim, contrary to some modern perspectives, that "the universe is beautified by sinners," but that only the godless fail to appreciate this point!

It is clear from the sharpness of various rebukes that Augustine regards judging the parts rather than the whole as not merely an error of judgment but a moral error. In his *Confessions*, he even associates this preference for the parts over the whole with pride and abandonment of God: "This is what happens, O Fountain of life, when we abandon you, who are the one true Creator of all that ever was or is, and each of us proudly sets his heart on some one part of your creation instead of on the whole."[99] Also in the *Confessions*, he rebukes his soul for delighting in the senses of the flesh, telling his soul that this delight "is only a part and you have no knowledge of the whole."[100] In these passages and some others, Augustine seems to associate the "whole" with God himself, since he is the fount of all being. This emphasis on the *completeness* of the whole is as important as his emphasis on its organization and beauty. Elsewhere, he compares the organization of the universe as a complete entity to particular objects within the universe that bear a similar organizational unity: "Every nature, whether it is perceived by merely sentient or fully rational observers, preserves, in its parts, being like one another, the effigy of the whole universe."[101]

Now, it must be clearly stated that Augustine's emphasis on the harmonization of evil in relation to the whole is not his *complete* answer to the problem of evil, but rather works in conjunction with his broader effort at theodicy. For instance, Augustine is famous for his "free will defense" against the problem of evil. In the *Enchiridion*,

---

[98]*De vera religione* 22.43 (CSEL 77, 30-31).
[99]*Confessiones* 3.8 (CSEL 33, 58).
[100]*Confessiones* 4.11 (CSEL 33, 77).
[101]*De Genesi ad litteram liber unus imperfectus* 15.59 (CSEL 28:1, 499).

he stipulates that "the cause of our evils is the will of a changeable good falling away from the unchangeable good, first the will of an angel, then the will of a human being."[102] Moreover, Augustine explicitly opposes the appeal to the "beauty of the whole" as a way to displace the notion of *fallenness*. For instance, in a lengthy intermezzo on the nature of pride in book 11 of his literal commentary on Genesis, he deals with the view that God made the devil bad from the beginning, drawn from a misuse of Job 40:19.[103] Those who advanced this argument, in order to account for the goodness of creation affirmed in Genesis 1:31, appealed to something like the problem of perspectival prejudice: those who are bad "do not succeed with their badness in disfiguring or upsetting at any point the beauty and order of the whole."[104] They also emphasized that their "deserts are weighed up" such that justice and the beauty of the world are maintained.[105] Augustine forcefully rejects this view, claiming that it is "manifestly contrary to justice" for God to condemn any creature for what he himself had created in them. Thus, it is evident that Augustine does not appear to conceive of the appeal to the "beauty of the whole" and a free will theodicy as mutually exclusive. In fact, an appeal to perspectival prejudice would appear problematic if isolated from a corollary emphasis on the misuse of free will by rational creatures.

A "greater good" theodicy also functions, in general terms, as a part of Augustine's resolution to the problem of evil. "God judged it better to bring good out of evil than not to allow evil to exist," he writes in the *Enchiridion*.[106] In his sermon expounding the Apostles' Creed referenced in chapter 1, Augustine emphasizes God's ability to use evil for good, providing the crucifixion of Christ as an example. God used it for great good, but it only came about through the malice

[102] *Enchiridion* 8.23, in *On Christian Belief*, 288.
[103] *De Genesi ad litteram* 11.21.28 (CSEL 28:1, 353-54).
[104] *De Genesi ad litteram* 11.21.28 (CSEL 28:1, 353).
[105] *De Genesi ad litteram* 11.21.28 (CSEL 28:1, 353).
[106] *Enchiridion* 8.27, in *On Christian Belief*, 290.

of the devil, the Jews, and Judas.[107] Augustine then declares: "So too in the hidden and secret recesses of the whole of creation, which neither our eyes nor our minds are sharp enough to penetrate, God knows how he makes good use of the bad, so that in everything that comes to be and is accomplished in the world the will of the Almighty may be fulfilled."[108] Augustine does not specify what he means here by "the hidden and secret recesses of the whole of creation," but the terms *secret* and *hidden* make it clear that he has some conception of God using evil for good throughout creation in ways that surpass human understanding.

CONCLUSION

Reference to these broader theodicy considerations allows us to conclude with one more limited point of application to the contemporary creation debate, and one broader consideration in relation to the problem of natural evil. First, we must appreciate that Augustine's treatment of this problem predates many of the scientific and philosophical challenges that mark the contemporary discussion. Augustine had no conception of a thirteen-billion-year-old universe or hundreds of millions of years of evolutionary struggle and competition. In fact, we have no reason to think that Augustine was particularly interested in the challenge of *pre-human* animal death. For this reason, it is all the more striking that Augustine, under the enormous pressure of Manichaean criticisms, does not respond, "Yes, animal death is bad, but this is only because Adam and Eve sinned." To be sure, the misuse of free will is a huge part of his broader theodicy (though here he puts more emphasis on the angelic fall than many contemporary views), but animal death itself is, for Augustine, not a problem to be solved so much as a beauty to be admired—a cause for praising God more than blaming him.

[107]Sermon 214.3, in *Sermons 184–229Z on the Liturgical Seasons*, 152.
[108]Sermon 214.3, in *Sermons 184–229Z on the Liturgical Seasons*, 152.

Now, of course, we might not agree with Augustine. For my own part, I consider Augustine's approach to the challenge of animal death most helpful when used in conjunction with other considerations in natural evil theodicy, and not an exhaustive and sufficient answer in itself. Nonetheless, many of his insights may be instructive at various points within the contemporary discussion. And at the very least, unless we are willing to question Augustine's orthodoxy, his views will certainly discourage us from accepting the claims of those who think that any acceptance of animal death prior to the human fall "undermines the very foundation of the gospel."[109]

Beyond this, and more basically, the developmental nature of Augustine's vision of created reality may enrich the nature of Christian hope in the face of the suffering and disorder we see in natural history, directing our attention forward rather than backward for the ultimate answer. Think of Augustine's claim that dying creatures "fill up the number of time-bound forms and shapes into a single beauty," or his metaphor of the universe as a successive poem, or his references to "the beauty of seasons" and God's "time-bound method of healing." This language, together with Augustine's reference to God's loftier and all-embracing purposes beyond human knowledge, may function not only to caution our judgment, but to invite our imagination. "Good but imperfect" need not be the whole story; perfection may be the goal rather than the starting point. God might be accomplishing something through the passage of time that is not yet visible.[110] Indeed, Augustine's language hints that the current state of the world may simply be one step—one line in a long poem—toward

[109]Avery Foley, "Did Adam Step on an Ant Before the Fall?," https://answersingenesis.org/death-before-sin/did-adam-step-on-an-ant-before-fall.

[110]E.g., cf. Philip Hefner: "Within the framework of Christian theology, the dynamic of nature is to be explained as its trajectory toward communion with God, according to God's intentions. . . . Nature is a realm of becoming" (as quoted in Ted Peters, "The Evolution of Evil," in *The Evolution of Evil*, 52).

something far better. Something of this hope can be captured in a final quote from the *Confessions*:

> It is always the case that the greater the joy, the greater is the pain which precedes it. . . . Why is it that in this part of your creation which we know there is this ebb and flow of progress and retreat, of hurt and reconciliation? Is this the rhythm of our world? Is this what you prescribed when from the heights of heaven to the depths of earth, from the first beginnings to the end of time, from the angel to the worm, from the first movement to the last, you allotted a proper place and a proper time to good things of every kind and to all your just works?[111]

[111]*Confessiones* 8.3 (CSEL 33, 176).

CHAPTER FIVE

CAN WE EVOLVE ON EVOLUTION WITHOUT FALLING FROM THE FALL?

Augustine on Adam and Eve

The account of [the fall of Adam and Eve] was written down here in the way in which it should be read by all, even if it would only be understood in the way in which it should be by a few.

De Genesi ad litteram 11.42.60

I opened the previous two chapters by observing how contentious Genesis 1 and animal death are in some evangelical circles. But others tend to be more relaxed in these areas. Many evangelicals are better represented by Ted Cabal's appeal that Christians should not divide over the age of the earth;[1] they tend to adopt a more irenic posture amidst differences in reading Genesis 1;[2] and

---

[1]Theodore J. Cabal and Peter J. Rasor, *Controversy of the Ages: Why Christians Should Not Divide Over the Age of the Earth* (Wooster, OH: Weaver, 2017), esp. 187-217.

[2]For instance, C. John Collins and R. Albert Mohler debated the question "Does Scripture Speak Definitively on the Age of the Universe?" at Trinity Evangelical Divinity School in February 2017, in connection with the Creation Project: http://henrycenter.tiu.edu/resource/genesis-the-age-of-the-earth-does-scripture-speak-definitively-on-the-age-of-the-universe. In this debate, Mohler adopted a more irenic approach than many other young-earth creationists advocate, casting the age of the earth as a third-rank issue on his "theological triage," and stipulating that he not only recognizes those who affirm an older earth and universe as fellow Christians—he hires them on his faculty.

they are often agnostic or openhanded about topics like prehuman animal death.[3]

But this does not mean that creation has ceased to be controversial. For much of evangelicalism, the controversy has simply shifted from the interpretation of the days of Genesis 1 (and the issues that broadly separate young-earth creationism from old-earth creationism) to the interpretation of Adam and Eve in Genesis 2–3 (and the issues that broadly separate both of these camps from evolutionary creationists). It is here, in issues related to Adam and Eve, that the deepest worry and energy seems to concentrate.[4] Even when evangelicals are debating other creation topics like the age of the world, Adam often elbows his way into view.[5]

In many respects, of course, worries about evolution are nothing new. The association of evolution with atheism and secularism has long been ingrained into evangelical consciousness. Historically, prominent evangelical leaders have claimed that "evolution was invented in order to kill the God of the Bible,"[6] and have blamed

[3]Thus, the 2017 Crossway book *Theistic Evolution* strongly criticizes theistic evolution but is explicitly openhanded about questions of the age of the world, and does not frame its concern along the lines of whether one interprets Genesis "literally." Similarly, Kenneth D. Keathley and Mark F. Rooker, in *40 Questions About Creation and Evolution* (Grand Rapids: Kregel Academic, 2014), survey various views with respect to the age of the earth and interpretations of Genesis 1 (237). At the same time, they stipulate, "We believe that the historicity of Adam and Eve is so important that the matter should serve as a litmus test when evaluating the attempts to integrate a proper understanding of Genesis 1–3 with the latest findings of science." Later, the authors provide a helpful threefold taxonomy of possible ways to affirm a historical Adam and Eve and some account of human evolution (379-83), all of which they regard as meeting the necessary criteria for establishing humanity's uniqueness, unity, and fallen condition (384).

[4]In their introduction to *Adam, the Fall, and Original Sin: Theological, Biblical, and Scientific Perspectives* (Grand Rapids: Baker Academic, 2014), editors Hans Madueme and Michael Reeves draw attention to various indicators of evangelical anxiety over Adam, such as the departures of prominent evangelical faculty members from Reformed Theological Seminary and controversy over positions taken by Francis Collins and Peter Enns (vii-x).

[5]It was interesting to observe how, in the Mohler-Collins debate just referenced, the issue of the historicity of Adam came up repeatedly as a topic of concern in spite of the debate's topic and the broad agreement among the participants on this issue. Nonetheless, as we will suggest, the participants' interest in Adam, and their more moderate claims about the age of the world, has a good claim to wisdom.

[6]John MacArthur, "Creation: Believe It or Not, Part 1," March 21, 1999, www.gty.org/library/sermons-library/90-208/creation-believe-it-or-not-part-1.

evolution for humanism, atheism, abortion, infanticide, euthanasia, pornography, adultery, divorce, homosexuality, premarital sex, the destruction of the family, animalism, promiscuity, vandalism, hedonism, cannibalism, and (if all that is not enough) Adolf Hitler.[7] Even among those evangelicals who are more openhanded about the age of the universe, it is not difficult to find assertions that the acceptance of evolution undermines the gospel itself.[8] All this is well familiar.

But now genetic findings have furthered the angst. In the eyes of many, any possible questions about the paleoanthropological data are now resolved beyond doubt by the claim from genetic science that the human population was never limited to two people.[9] Adam and Eve are thus increasingly in the spotlight—and a good case can be made that they deserve it. The theological stakes are high. Among the most important questions are:

1. Were Adam and Eve historical individuals?
2. Were Adam and Eve fresh, de novo creations, or descended from previous hominins?[10]
3. Were Adam and Eve the first human beings, and the only living human beings at the time of their creation?
4. Were Adam and Eve the biological progenitors of all living human beings?

---

[7]David Jeremiah, "Foreword," in Henry M. Morris, *The Long War Against God: The History and Impact of the Creation/Evolution Conflict* (Grand Rapids: Baker, 1989), 10.

[8]Wayne Grudem, "Foreword," in *Should Christians Embrace Evolution? Biblical and Scientific Perspectives*, ed. Norman C. Nevin (Phillipsburg, NJ: P&R, 2009), 10.

[9]For an overview of the relevant science, see Dennis R. Venema and Scot McKnight, *Adam and the Genome: Reading Scripture After Genetic Science* (Grand Rapids: Brazos, 2017), 1-91.

[10] *Hominins* is now generally taken as the better term in this context, referring to all modern and extinct humans and their immediate ancestors, while the term *hominids* refers more broadly to all modern and extinct Great Apes. The terminology is not used consistently in this way, however. For a discussion of the relevant terminology, and an overview of different positions on hominid fossils, see the discussion in *Dictionary of Christianity and Science: The Definitive Reference for the Intersection of Christian Faith and Contemporary Science*, ed. Paul Copan, Tremper Longman III, et al. (Grand Rapids: Zondervan, 2017), 355-65.

It is not difficult to see that such questions have massive theological repercussions, most immediately for the doctrines of the fall and original sin. What we affirm about Adam and Eve is foundational to our view of the human need for salvation, and in this sense foundational to the gospel. The age of the universe, and animal death, have no comparable reach.

In this chapter, we will draw Augustine into this area of evangelical debate and anxiety. Obviously, Augustine did not face this discussion as we do, and he cannot resolve all of the questions we have. Actually, he will introduce new complexities of his own. Nonetheless, as we will see, his view of Adam and Eve is highly relevant to many contested points in the current dialogue, particularly to the extent that we are willing to receive his influence on attitudinal (not just material) issues. And even where Augustine does not provide an answer, he may still encourage us toward a greater recognition of the complexity of the issue, which is itself often a helpful ingredient in the pursuit of truth.[11]

First, we will consider the common claim that Augustine's view of *rationes seminales* is favorable to the theory of biological evolution. Then we will survey Augustine's views on the creation, historicity, and nature of Adam and Eve, particularly as set forth in his literal commentary on Genesis. Finally, we will bring Augustine's views into dialogue with the contemporary scene, identifying several "instincts" represented among evangelical views of Adam and Eve, and suggesting possible pathways forward that Augustine might encourage.

[11]There is a principle of textual criticism labeled *lectio difficilior potior*: "The more difficult reading is the stronger." That is, all other factors being equal, a scribe is more likely to smooth out a textual difficulty than to create one, and therefore in manuscript variations it is often the more unexpected reading that is the original one. Might there be something analogous to this principle in our pursuit of truth wholesale? That is, all other factors being equal, it is the more complicated version of truth that tends to be right, since it is less likely to be invented or discovered unless necessary. Such an intuition will doubtless fail to be consistently reliable, but it may at least incline us to be open to considerations of complexity, nuance, and synthesis.

IS AUGUSTINE'S NOTION OF "SEMINAL REASONS" FAVORABLE TO BIOLOGICAL EVOLUTION?

Augustine and evolution have long been seen as well-suited allies. As early as 1896, the distinguished Catholic physicist at the University of Notre Dame, John Augustine Zahm, claimed that "it was the great bishop of Hippo who first laid down the principles of theistic Evolution as they are held to-day."[12] Nearly a century ago, Henry Woods noted this tendency among Catholic theologians to garner support for the theory of evolution by appealing to Augustine's principle of *rationes seminales* ("seminal reasons").[13] The question was not, of course, whether *rationes seminales* is identical to evolution, but whether the principles entailed by this aspect of Augustine's thought are favorable to it.[14] At the outset of his discussion, Woods cautioned against the danger of reading our own ideas back into Augustine: "Those who would draw from St. Augustine support for Evolution, must take his doctrine in his own sense simply and without gloss, neither grudging the toil without which it can not [*sic*] be penetrated, nor yielding to the temptation to read into it their own sense."[15] Woods went on to argue vigorously that no connection can be made between *rationes seminales* and the theory of evolution.[16]

All these years later, a connection between these two ideas is still often advanced. Alister McGrath, for instance, advises a careful use of Augustine's idea:

> Neither Augustine nor his age believed in the evolution of species. There were no reasons at that time for anyone to believe in this notion. Yet Augustine developed a theological framework that could

[12]Quoted in Paul M. Blowers, *Drama of the Divine Economy: Creator and Creation in Early Christian Theology and Piety*, Oxford Early Christian Studies (Oxford: Oxford University Press, 2012), 157.
[13]Henry Woods, *Augustine and Evolution: A Study in the Saint's* De Genesi ad litteram *and* De Trinitate (New York: The Universal Knowledge Foundation, 1924).
[14]Woods, *Augustine and Evolution*, 3.
[15]Woods, *Augustine and Evolution*, 3-4.
[16]Woods, *Augustine and Evolution*, 46-48; cf. 143-48.

> accommodate this later scientific development, though his theological commitments would prevent him from accepting any idea of the development of the universe as a random or lawless process.[17]

Is this a valid appeal from the notion of *rationes seminales*? To answer this, we must consider what exactly Augustine meant by this term. In its broad form, it is not original to Augustine but derives from Stoic and Platonic philosophy. At its core it means that, as McGrath summarizes, "God created the world complete with a series of dormant multiple potencies, which were actualized in the future through divine providence."[18] Throughout his literal commentary, Augustine uses the terms *reasons* or *ideas* (*rationes*),[19] as well as *seeds* (*semina*),[20] to refer to that which God embedded into the world in the moment of its instantaneous creation. In one passage, he describes the entire created order carrying on its movement—angels carrying out God's orders, constellations circling around their courses, wind and water moving, plants breeding and evolving, animals living their lives, and the wicked vexing the just—as the development of all those principles God instilled into the world in the initial creative moment. As he puts it, God "unwinds the ages which he had as it were folded into the universe when it was first set up."[21] In another passage, he uses the image of a tree growing from a seed.[22] He asks his readers to picture a tree, and reminds them that the trunk, branches, leaves, and fruits it possesses did not all spring into being at once, but rather in a long process from a seed. He then argues for a similar vision of the entire universe:

---

[17]Alister McGrath, *The Passionate Intellect: Christian Faith and the Discipleship of the Mind* (Downers Grove, IL: InterVarsity Press, 2010), 145. Cf. also McGrath, *A Fine-Tuned Universe: The Quest for God in Science and Theology* (Louisville, KY: Westminster John Knox, 2009), 101-6.

[18]McGrath, *A Fine-Tuned Universe*, 103.

[19]*De Genesi ad litteram* 4.33.51 (CSEL 28:1, 129).

[20]*De Genesi ad litteram* 5.7.20 (CSEL 28:1, 150).

[21]*De Genesi ad litteram* 5.20.41 (CSEL 28:1, 164-65).

[22]*De Genesi ad litteram* 5.23.45 (CSEL 28:1, 168); cf. also *De Genesi contra Manichaeos* 1.6.10 (CSEL 91, 77).

> Just as all these elements, which in the course of time and in due order would constitute a tree, were all invisibly and simultaneously present in that grain, so too that is how, when God created all things simultaneously, the actual cosmos is to be thought of as having had simultaneously all the things that were made in it and with it *when the day was made* (Gen 2:4).[23]

Augustine's vision of developmental methods of creation has obvious similarities with the notion of evolution broadly, but it must not simply be equated with the modern theory of biological evolution specifically. The notion that God stopped creating new species is Augustine's solution to the apparent conflict between God's rest on the seventh day of Genesis 1 and Jesus' assertion in John 5:17 that "my Father is working until now."[24] In fact, Augustine affirms that every animal species was created in the initial creation event such that the subsequent emergence of any new species would contradict the biblical affirmation of God's rest after his completed work. As he puts it, "But clearly, if we suppose that he now sets any creature in place in such a way that he did not insert the kind of thing it is into that first construction of his, we are openly contradicting what scripture says, that he finished and completed all his works on the sixth day."[25] Since Augustine affirmed the fixity of species, we must consider the *principles* involved in the notion of *rationes seminales* in relation to biological evolution, rather than draw any direct correlation between the two.

It is significant that Augustine envisions the unfolding or "unwinding" of the *rationes seminales* to involve both natural processes and supernatural intervention.[26] The seminal reasons anticipate, he

---

[23]*De Genesi ad litteram* 5.23.45 (CSEL 28:1, 168).

[24]*De Genesi ad litteram* 4.12.23 (CSEL 28:1, 108-10).

[25]*De Genesi ad litteram* 5.20.41 (CSEL 28:1, 164).

[26]I am aware that the distinction between "natural" and "supernatural" events is disputed, but I believe this terminology is useful and can be defended. For some discussion, see C. John Collins, *Did Adam and Eve Really Exist? Who They Were and Why You Should Care* (Wheaton, IL: Crossway, 2011), 108; and Collins, *The God of Miracles: An Exegetical Examination of God's Action in the World* (Wheaton, IL: Crossway, 2000).

argues, both those "temporal events [that] most commonly transpire" (for instance, living creatures created slowly through reproductive processes), as well as the fact that "rare and miraculous things are done" (as Augustine conceives the creation of the body of Adam).[27] He also asserts God's ability to intervene and redirect the unfolding of the seminal reasons: "Over and above this natural course and operation of things, the power of the creator has in itself the capacity to make from all these things something other than what their seminal formulae, so to say, prescribe."[28] Thus, Augustine does not view the unfolding of the seminal reasons as an inflexible, impenetrable process that is impervious to external redirection.

Related to this, the unfolding of the seminal reasons should not be understood in a way that is compatible with deism: as an autonomous unfolding of prior principles from which God has stepped back and is no longer involved. Rather, Augustine thinks that God creates the world with a built-in capacity to develop, and that this subsequent development is entirely harmonious with God's intervention and involvement. This is a point of difference with Plotinus's more naturalistic construal *rationes causales*—for Augustine, the reasons determine a variety of eventualities, and the subsequent arrival at one possibility over another is determined by the will of God rather than being inherent in, and determined by, the original reason.[29]

Thus, in my view, Augustine's principle of *rationes seminales* is underdetermined with respect to any wholesale judgment about evolution. The wisest and most responsible appeal to this principle will be cautious, not least in light of the differences involved in Augustine's context and ours—for instance, Augustine would not have anticipated that *rationes seminales* would be employed in favor of views that undermine the

[27]*De Genesi ad litteram* 6.14.25 (CSEL 28:1, 189).

[28]*De Genesi ad litteram* 9.17.32 (CSEL 28:1, 291).

[29]This point is well made by Rowan Williams, "Creation," in *Augustine Through the Ages: An Encyclopedia*, ed. Allan D. Fitzgerald (Grand Rapids: Eerdmans, 1999), 252.

historicity of the fall. While there are obvious points of resonance between *rationes seminales* and the theory of evolution, we should restrain our definite conclusions to the following two points.

First, Augustine's principle of *rationes seminales* will undermine the intuition that de novo processes of creation are inherently superior to those involving intermediate causes and natural processes. In other words, if and when God creates through evolutionary processes, this in no way involves a necessary diminution of his goodness, wisdom, and power. This might appear to be a fairly basic point, but it is relevant to the current debate. For instance, some opposition to evolutionary creationism seems to involve the assumption that direct creation is linked with God's power, while creation through gradual process involving natural causation is linked with deism. Wayne Grudem, for example, describes theistic evolutionists as those who surrender to evolution and then "tack on God, not as the omnipotent God who in his infinite wisdom directly created all living things, but as the invisible deity who makes absolutely no detectable difference in the nature of living beings as they exist today."[30]

One need not be committed to an evolutionary account of creation, however, to wonder about the function of the words *omnipotent*, *directly*, *wisdom*, and *invisible* in this sentence. Why should direct forms of creation display God's wisdom and omnipotence, but indirect forms of creation imply divine invisibility and a kind of deism? In the Bible, for instance, natural processes in which there is no detectable supernatural activity are attributed precisely to God's power and wisdom, such as the formation of clouds, rain, and wind (Ps 135:7), the growing of grass (Ps 147:8), and the development of life in the womb (Ps 139:13).[31] Take the existence of a particular human being

[30]Grudem, "Foreword," in *Should Christians Embrace Evolution?*, 10.

[31]For further elaboration of this point, see Denis R. Alexander, "Creation, Providence, and Evolution," in *Knowing Creation: Perspectives from Theology, Philosophy, and Science* (Grand Rapids: Zondervan, 2018), 284-85.

alive today. Can we not declare that our existence is as much a manifestation of God's power and wisdom as any direct action of creation, even though it involved a sperm fertilizing an egg in our mother's fallopian tubes, and the entire subsequent "natural" process that followed? The Bible describes this development from a zygote to an embryo to a fetus as God "knitting." This is not an argument for evolution wholesale; it is simply the observation that God can create through natural processes. It is not an inferior way to create. This is consistent with Augustine's idea of *rationes seminales*; in fact, as we will see, Augustine makes this exact point about God being equally the creator of Adam and every subsequent human being, despite the fact that he thinks Adam's creation occurred more directly than other human beings are created.

Second, in the other direction, Augustine's principle of *rationes seminales* also undermines the intuition that de novo mechanisms of creation are inherently inferior to those involving intermediate causes and natural processes. In other words, if God were to intervene supernaturally within a natural process, this would in no way imply a blemish or oversight in the original design. Sometimes voices in the evolutionary creationist camp give the impression that divine intrusion in the creative process in discernible ways is a kind of imperfection, or at least an oddity.[32] Augustine would not sympathize with

[32]For instance, Francis S. Collins, *The Language of God: A Scientist Presents Evidence for Belief* (New York: Free Press, 2006), 193-94: "[Intelligent design] portrays the Almighty as a clumsy Creator, having to intervene at regular intervals to fix the inadequacies of His own initial plan for generating the complexity of life." While I appreciate much of Collins's work, it is not clear that it is necessary to conceive of divine intervention in the evolutionary process as a matter of "fixing inadequacies" or a sign of divine clumsiness. Similarly, Howard Van J. Till, "Intelligent Design: The Celebration of Gifts Withheld?," in *Darwinism Defeated? The Johnson-Lamoureux Debate on Biological Origins* (Vancouver: Regent College Publishing, 1999), 86-87, describes progressive creationist views as requiring a God who withholds "formational gifts" from it, and instead invests it with "gaps in the Creation's formational economy" that constitute an incompleteness and seeming imperfection in the design. It is not for nothing that respondents to Van Till's portrayal of theistic evolution struggle to disassociate it from a kind of deistic instinct. See Walter L. Bradley, "Response to Howard J. Van Till," in *Three Views on Creation and Evolution*, ed. J. P. Moreland and John Mark Reynolds, Counterpoints (Grand Rapids: Zondervan, 1999), 224; John Jefferson Davis, "Response to Howard J. Van Till," in *Three Views on Creation and*

this intuition. As we have noted, Augustine understands the unfolding of the seminal reasons as involving various kinds of divine activity, including what we call "miracles." As Simo Knuuttila summarizes Augustine's thought on this point, "Miraculous events are not unnatural. Our concept of nature is based on observational regularities, but the ultimate nature is God's will or providential design, which provides natural history with all kinds of exceptional events."[33] Similarly, Rowan Williams, developing Augustine's doctrine of *rationes causales*, comments, "a 'miraculous' event does not offend against the integrity of the created order: every reality from the first contains the possibility of whatever may subsequently happen to it, so that it will not do to think of a miracle as *contra naturam*."[34]

Thus, Augustine's vision of God implanting *rationes seminales* into the created order at the instant of its creation should not determine, in and of itself, our stance toward the theory of biological evolution. It should, at least, discourage us from dismissing evolutionary processes of creation a priori, as though God must always create quickly and directly, rather than slowly and indirectly, for it to be really a work worthy of divine power. In the other direction, the possibility of secondary causes and gradual processes does not require them, and we should resist any framework that excludes supernatural intervention into natural processes at the outset, as though such an intervention inevitably and necessarily falls prey to the "God of the gaps" fallacy.[35]

To put it in colloquial terms, creation via evolution is not a second-rate way of getting the job done, but neither does supernatural intervention

---

Evolution, 228; and Vern S. Poythress, "Response to Howard J. Van Till," in *Three Views on Creation and Evolution*, 237-38.

[33]Simo Knuuttila, "Time and Creation in Augustine," in *The Cambridge Companion to Augustine*, ed. David Vincent Meconi and Eleonore Stump, 2nd ed. (Cambridge: Cambridge University Press, 2014), 86.

[34]Williams, "Creation," in *Augustine Through the Ages*, 252.

[35]For further argumentation that not all appeals to special divine action are prey to the "God of the gaps" objection, see C. John Collins, "How to Think about God's Action in the World," in *Theistic Evolution*, 659-81.

imply a correction to an earlier mistake. God might simply create different things differently. We cannot absolutely decide in advance. Thus, while it does indeed have a striking resonance with broadly evolutionary ways of understanding God's work of creation, Augustine's principle of *rationes seminales* ultimately leaves us open to follow the data, both biblical and scientific, and to evaluate the theory of biological evolution on its evidential merits.

AUGUSTINE ON THE CREATION OF ADAM AND EVE

In his allegorical commentary, Augustine favors a figurative interpretation of the creation of Adam and Eve, but insists a literal one is possible as well. Referencing God's creation of Adam "from the mud of the earth" in Genesis 2:7, Augustine inquires "what sort of mud this was, or what material was being signified by the word 'mud.'"[36] Like the Vulgate, Augustine uses the Latin term *limus* for "mud" here. Augustine apparently considers the question of what is signified by this "mud" to be a commonly addressed issue, calling it the "first question . . . that is usually raised" about this verse.[37] In particular, Augustine here references "these enemies . . . of the books of the Old Testament, looking at everything in a fleshly, literal-minded way"—that is, the Manichaeans—who wonder why God did not create the man out of better material than mud.[38] Augustine counters by gesturing toward the multifarious meaning of "mud" as a mixture of both water and earth: "What they fail to understand from the start is how many meanings both earth and water are given in the scriptures—mud, you see, being a mixture of water and earth."[39] He proceeds to argue that even if God did make Adam from the mud of this earth, there is no reason why he could not have given him a body that would have been incorruptible apart from sin.[40]

---

[36]*De Genesi contra Manichaeos* 2.7.8 (CSEL 91, 127).
[37]*De Genesi contra Manichaeos* 2.7.8 (CSEL 91, 127).
[38]*De Genesi contra Manichaeos* 2.7.8 (CSEL 91, 127).
[39]*De Genesi contra Manichaeos* 2.7.8 (CSEL 91, 127).
[40]*De Genesi contra Manichaeos* 2.7.8 (CSEL 91, 127-28).

Augustine displays a similar latitude as he addresses the question of the creation of Eve. He defends the possibility of a literal creation from Adam's rib, but displays more interest in the "hidden" meaning of the event: "Even if the real, visible woman was made, historically speaking, from the body of the first man by the Lord God, it was surely not without reason that she was made like that—it must have been to suggest some hidden truth."[41] Ultimately, Augustine interprets Eve's creation in connection to the sacraments of the church and the relationship of a husband and wife, urging that while a literal interpretation of Eve from Adam's side is possible, it is not the point of the text: "Was there any shortage of mud, after all, for the woman to be formed from? . . . So whether all this was said in a figurative way, or whether it was even done in a figurative way, it was certainly not pointlessly that it was said or done like this. No, it is all assuredly pointing to mysteries and sacraments."[42] Although Augustine's comments here come in a different context, we already see the balanced set of concerns that we will note throughout this chapter—on the one hand, his desire to resist the complete allegorizing of the Adam and Eve story in Genesis 2–3; and on the other, his opposition to overliteralism in the process. Here and elsewhere, however, he will tolerate symbolical readings as orthodox, while he regards the "fleshly" literalism of Manichaeism as an assault on the Christian faith.

Now, even if Augustine had confined such views to his allegorical commentary, they might still be significant in drawing attention to how Genesis 2–3 was read in his time and what interpretations he himself was willing to countenance. But strikingly, when he gets to his literal commentary, Augustine retains much of his flexibility on the question of the creation of Adam and Eve. Here, though, Augustine's notion of instantaneous creation poses a challenge. Is the creation of the man Adam from the dust in Genesis 2:7 a recapitulation

---

[41]*De Genesi contra Manichaeos* 2.12.17 (CSEL 91, 127).
[42]*De Genesi contra Manichaeos* 2.12.17 (CSEL 91, 127).

of the creation of humanity in Genesis 1:26-28? If so, does Genesis 2:7 also belong to the initial instantaneous creative event? Augustine ponders these questions in book 6 of his literal commentary. Ultimately, he affirms that the creative work in view in Genesis 2:7, along with the creation of Eve from Adam's rib in Genesis 2:22, belongs to God's creative work in time, subsequent to the initial simultaneous creation.[43] The creation of humanity in Genesis 1:26-28 must therefore be understood "in terms of a potentiality inserted as it were seminally into the universe through the Word of God when he created all things simultaneously."[44]

Thus, Augustine thinks that Genesis 1 narrates the creation of humanity invisibly, potentially—while Genesis 2 narrates the creation of humanity visibly, as we now know human constitution.[45] Augustine distinguishes this twofold means of creation from his idea of *rationes seminales*; he argues that the creation of humanity in Genesis 1 was not even yet in seed form, but merely causal.[46] Later, he says the "idea" (*ratio*) of man was made in Genesis 1, not the actual man.[47] He acknowledges the difficulty of this view, but insists that it is necessary in light of Genesis 2:4 and Sirach 18:1, texts he understands to require an initial instantaneous creation.[48]

In his literal commentary, Augustine regards the creation of Adam as a supernatural, de novo event. He conceives of Adam as the first man, and denies that he had parents.[49] Similarly, Augustine will affirm that Eve was truly made from Adam's body, though he emphasizes the mystical symbolism of this event and the possible agency of angels in the removal of Adam's rib during sleep.[50] At the same time,

---

[43]*De Genesi ad litteram* 6.2.3-3.5 (CSEL 28:1, 171-73).
[44]*De Genesi ad litteram* 6.5.8 (CSEL 28:1, 176).
[45]*De Genesi ad litteram* 6.6.10 (CSEL 28:1, 177).
[46]*De Genesi ad litteram* 6.6.10-11 (CSEL 28:1, 177-78).
[47]*De Genesi ad litteram* 6.9.16 (CSEL 28:1, 182).
[48]*De Genesi ad litteram* 6.6.11 (CSEL 28:1, 178).
[49]*De Genesi ad litteram* 6.15.26 (CSEL 28:1, 189).
[50]*De Genesi ad litteram* 9.13.23-9.16.30 (CSEL 28:1, 284-90).

he cautions against a literalistic understanding of the biblical language used to describe these events. He first warns against any notion of God being *physically* involved in Adam's creation, insisting that the text's reference to God *fashioning* Adam is metaphorical.[51]

Then Augustine tackles the nature of the event itself: "But in what manner did God make him from the mud of the earth? Was it straightaway as an adult, that is, as a young man in the prime of life? Or was it as he forms human beings from then until now in their mothers' wombs?"[52] Augustine ultimately leaves this question open, wondering whether it is our business to even ask about, and asserting that "whichever of these he did, after all, he did what was in the power of a God both omnipotent and wise, and what befitted him to do."[53] He proceeds to appeal to the *rationes seminales* to bulwark his openness on this question, asserting that when things "develop from a latent to a manifest state of being," this in no way entails that God "abdicated the supremacy over everything of his will."[54] Augustine appeals to Jeremiah 1:5 ("Before I formed you in the womb, I knew you") for support of this point. In his sermons as well, Augustine quotes Jeremiah 1:5 as evidence that although all men come from the first man God made, God is no less the Creator of them than he was of the first man: "So at the beginning he created man without man, now he creates man from man. Still, whether it's man without man, or man from man, it is *he who made us, and not we ourselves* (Ps 100:3)."[55] Here, as we emphasized just above, Augustine is eager to assert that God's use of secondary means detracts nothing from the divine wisdom and power displayed in his creative work.

---

[51]*De Genesi ad litteram* 6.12.20 (CSEL 28:1, 185).

[52]*De Genesi ad litteram* 6.13.23 (CSEL 28:1, 187).

[53]*De Genesi ad litteram* 6.13.23 (CSEL 28:1, 187-88).

[54]*De Genesi ad litteram* 6.13.23 (CSEL 28:1, 188).

[55]Sermon 26.1, in *Sermons 20–50 on the Old Testament*, ed. John E. Rotelle, trans. Edmund Hill, The Works of Saint Augustine: A Translation for the 21st Century (Hyde Park, NY: New City Press, 1992), 93.

Augustine's flexibility on the question of Adam's age at the time of his creation is undergirded by his emphasis on God's sovereign direction over natural processes. One can understand why evolutionary creationists see Augustine's thought on this point as favorable to their view. To be sure, the differences should not be downplayed. Augustine resists allegorizing Genesis 2–3. At the same time, as he does so the openness of his *posture* is remarkable. Consider the speculations of his allegorical commentary, or his willingness to countenance the creation of an infant Adam slowly maturing through childhood and adolescence (what Augustine calls "the ages of man"). While Augustine is a resource to those who want a historical Adam, it would be misleading to prop him up as a tool to this end without appreciating the nuances in his view of Adam.

AUGUSTINE ON THE HISTORICITY OF ADAM AND THE GARDEN OF EDEN

A similar posture characterizes Augustine's treatment of the historicity of Adam and the garden of Eden. In his allegorical commentary on Genesis, he interprets Eden symbolically, seeing it as a figurative display of the "spiritual delights which go with the life of bliss." Thus, the many trees in the garden are spiritual joys, the tree of life is wisdom, the tree of the knowledge of good and evil is the "halfway centrality of the soul," the four rivers are the virtues of prudence, fortitude, temperance, and justice, and so forth.[56] In this context, Augustine interprets other events in the story, such as the serpent's appearance and God's walking in the garden, in terms of every human's experience of temptation and sin, reasoning that "the reason all this is written down, after all, is to put us on our guard against such things at the present time."[57] Various other elements of the story

[56]*De Genesi contra Manichaeos* 2.9.12-10.13 (CSEL 91, 131-34).

[57]*De Genesi contra Manichaeos* 2.15.22 (CSEL 91, 143); in the larger context of *De Genesi contra Manichaeos* 2.14.20-16.24 (CSEL 91, 141-47).

are interpreted symbolically as well—for instance, the angel with the flaming sword represents "temporal punishments and pains."[58]

Strikingly, Augustine describes all this as the "historical" meaning of the text, and so he then proceeds to develop its "prophetical" meaning in terms of Christ and the church: "But I promised that in this book I would consider first the account of things that have happened, which I think has now been unfolded, and go on to consider next what they prophesy."[59] Then Augustine proceeds to interpret the whole story prophetically in terms of Christ and the church (of course, he does not hesitate to draw attention between the serpent and the Manichaeans at this point).[60] That Augustine can consider the text's historical meaning to be concerned with all these various symbolical interpretations he has provided once again demonstrates the extent to which he does not regard "literal" interpretation to be devoid of allegory or figurative language, as noted in chapter 3.

In his literal commentary, Augustine once again adjusts his strategy. Here he takes up the question of whether the garden of Eden should be understood as a real historical-geographical place. His posture of both groundedness and flexibility on this point is evident in book 6, where he poses the question of whether Adam, while in the paradise of Eden, had the kind of body we shall possess in the resurrection.[61] Augustine expresses his discomfort with this possibility, since he wants to take the trees and fruits mentioned in Genesis 2–3 in a historical sense, and he regards this as difficult to square with Adam possessing a fully glorified body.[62] At the same time, he offers the concession that "if no other solution can be found it is better that we

---

[58] *De Genesi contra Manichaeos* 2.23.35 (CSEL 91, 158-59).

[59] *De Genesi contra Manichaeos* 2.24.37 (CSEL 91, 160). Augustine repeats this view that the bulk of his allegorical commentary has been concerned with "historical signification" in *De Genesi contra Manichaeos* 2.27.41 (CSEL 91, 167).

[60] *De Genesi contra Manichaeos* 2.24.37-26.40 (CSEL 91, 160-66).

[61] *De Genesi ad litteram* 6.20.31 (CSEL 28:1, 194).

[62] *De Genesi ad litteram* 6.21.32 (CSEL 28:1, 195).

should opt for understanding Paradise in a spiritual sense."[63] Ultimately, Augustine concludes here that it is not necessary to interpret Eden symbolically, reasoning that Adam could have received a transformed body if he had refrained from sin during his probationary period in Eden.[64] Nonetheless, it is also clear that Augustine does not regard a symbolical interpretation of Eden as a breach of orthodoxy.

But at the start of book 8 of the literal commentary, Augustine addresses the question of Eden's historicity at greater length, touching also on Adam's historicity. He opens this chapter by quoting Genesis 2:8 ("And God planted a paradise in Eden to the East, and put there the man whom he had fashioned"), and proceeds by acknowledging his awareness of the wide range of opinions about the nature of the paradise mentioned in this verse: "I am not ignorant that many people have said many things about this paradise."[65] In particular, he identifies "three generally held opinions" about this issue: first, those who interpret Eden "bodily" (*corporaliter*), as a real physical environment; second, those who interpret Eden "spiritually" (*spiritaliter*), as a symbol or type; and third, those who interpret Eden in both senses, as a real physical environment that simultaneously symbolizes a spiritual reality.[66] He then specifies that he favors the third option for Eden as well as Adam. Thus, although Adam functions in a typological and figurative sense, as the apostle Paul teaches, nonetheless he can still be understood as a real human being who lived a certain number of years and had a certain number of progeny.[67] In the same way, Eden can be understood as both a type of spiritual paradise and also as "quite simply a particular place on earth, where the man of earth would live."[68]

---

[63]*De Genesi ad litteram* 6.21.32 (CSEL 28:1, 195).
[64]*De Genesi ad litteram* 6.23.34 (CSEL 28:1, 196).
[65]*De Genesi ad litteram* 8.1.1 (CSEL 28:1, 229), my translation.
[66]*De Genesi ad litteram* 8.1.1 (CSEL 28:1, 229).
[67]*De Genesi ad litteram* 8.1.1 (CSEL 28:1, 229). Augustine uses the Latin term *forma* in discussing Adam as a type.
[68]*De Genesi ad litteram* 8.1.1 (CSEL 28:1, 229).

It is noteworthy that Augustine shows awareness of symbolical/typological readings of Genesis 2–3. He was obviously familiar with Ambrose's allegorical preaching on Genesis, but at this point he may be thinking specifically of Origen's view. Origen stood out not only for his odd speculation about preexistent souls being infused into human bodies, but more basically for his affirmation of a symbolical interpretation of Genesis 1-3:

> What man of intelligence, I ask, will consider it a reasonable statement that the first and the second and the third day, in which there are said to be both morning and evening, existed without sun and moon and stars, while the first day was even without a heaven? And who could be found so silly as to believe that God, after the manner of a farmer, "planted trees eastward in Eden," and set therein a "tree of life," that is, a visible and palpable tree of wood, of such a sort that anyone who ate of this tree with bodily teeth would gain life; and again that anyone who ate of another tree would get a knowledge of "good and evil"? And further, when God is said to "walk in the paradise in the evening" and Adam to hide himself behind a tree, I do not think anyone will doubt that these statements are made in scripture in a figurative manner, in order that through them certain mystical truths may be indicated.[69]

Here Origen apparently regards the notion of ordinary days in Genesis 1 to be as troubling as the notion of an ordinary garden in Genesis 2, instead urging a figurative interpretation as the only one worthy of intelligent interpreters.

Parallels between Augustine's exegesis of Genesis 1–3 and many of Origen's ideas have often been noticed, though they are often accounted for by intermediate texts, such as Ambrose's homilies on Genesis 1, Hilary's treatises on the Psalms, and Gregory of Elvira's treatise on the creation of man.[70] György Heidl, however, has argued

---

[69]Origen, *On First Principles* 4.3.1, trans. G.W. Butterworth (New York: Harper & Row, 1966), 288-89.

[70]György Heidl, *The Influence of Origen on the Young Augustine: A Chapter in the History of Origenism*, Gorgias Eastern Christian Studies 19 (Piscataway, NJ: Gorgias, 2009), 78.

for a direct influence of Origen's view on Augustine, even as early as prior to Augustine's writing of his allegorical work against the Manichaeans, through a Latin compilation of Origen's views circulating in the fourth century.[71] At the same time, if Augustine has Origen in mind with the first of the three views of the interpretation of Eden he mentions, it is not necessary to assume that he would have associated this view *exclusively* with Origen. In the generation immediately preceding Augustine, for instance, Julian the Apostate had posited that humanity derived from a plurality of couples, and Gregory of Nyssa had suggested that the body of Adam was constructed from animal ancestors.[72]

It is sometimes assumed that premodern interpreters of Genesis 2–3 always understood the text as a straightforward, photographic, historical narration, more similar than dissimilar to modern historiography. But in fact, readers well before Darwin appreciated the rich imagery and stylized nature involved in the story of the garden of Eden.[73] Throughout the history of Jewish and early Christian interpretation, the symbolism of the story was particularly emphasized, with Adam often functioning as a type of Israel, Eve a type of Mary, Eden a type of Sinai, and banishment from Eden as a type of exile to Babylon.[74] However, it must be equally appreciated that premodern exegetes did not generally set these literary features as at odds with

[71]Heidl, *Influence of Origen*, 79-163.

[72]As noted by David Livingstone, *Adam's Ancestors: Race, Religion, and the Politics of Human Origins* (Baltimore: Johns Hopkins University Press, 2008), 6-7.

[73]Philo, for instance, spoke of Genesis as containing "mythical fictions" that served as allegorical symbols to make ideas visible. See the discussion in Venema and McKnight, *Adam and the Genome*, 160.

[74]As an example, consider the parallels drawn in this Jewish midrash in the Genesis Rabbah (AD 300–500), in which God declares, "Just as I led Adam into the Garden of Eden and gave him a commandment and he transgressed it, whereupon I punished him with dismissal and expulsion and bewailed him by crying *ekah*—how could this be—so for the nation of Israel. . . . And just as I brought Adam and Eve into Eden so I brought my people into the land of Israel and gave them commandments. [Like Adam and Eve] they too transgressed my commandments and I punished them with dismissal and expulsion and bewailed them crying *ekah* [Lam. 1:1], which means "how could this be" (*Genesis Rabbah* 19.9, as quoted in Gary A. Anderson, *The Genesis of Perfection: Adam and Eve in Jewish and Christian Imagination* [Louisville, KY: Westminster John Knox, 2001], 15).

historicity. Genesis 2–3 was overwhelmingly regarded as providing the history of the first two human beings, whose sin brought death and ruin to all their biological descendants—even if readers appreciated subtleties in how it did so. The relative rarity of Origen's view, and the differences between Augustine and Origen, must not be diminished. While Augustine detected a symbolical thrust to Genesis 2–3, he did not claim with Origen that it is "silly" to believe in an actual "bodily" tree that imparts life or death when you eat its fruit.

Perhaps more useful than identifying Augustine's view, however, is seeing how he arrives at it—for instance, in this case, how he interacts with and ultimately categorizes symbolical readings of Adam and Eden. In his commentary, Augustine raises a number of considerations as he urges against reducing the story of Genesis 2–3 to allegory. While he encourages the pursuit of the text's figurative meaning, he stipulates that we should *first* seek out its historical meaning, and he describes the genre of Genesis 2–3 as closer to the biblical book of Kings than to Song of Solomon.[75] Furthermore, the fact that this story involves some things with which we are not familiar should not cause us to suspect its historicity, since it involves pivotal events that come first in the history of humanity.[76]

Augustine then proceeds, however, to argue that although he favors viewing Adam and Eden as both historical and typological, there are many who accept the authority of the Scriptures who regard Eden as a type: "I am addressing, of course, those who accept the authority of these sacred writings; some of them, you see, are not prepared to have Paradise understood literally and properly, but only figuratively."[77] Augustine then distinguishes the view of these interpreters from the more insidious position of the Manichaeans: "As for those who are altogether opposed to these writings, I have dealt with them elsewhere

[75]*De Genesi ad litteram* 8.1.2 (CSEL 28:1, 229-30).
[76]*De Genesi ad litteram* 8.1.2 (CSEL 28:1, 230).
[77]*De Genesi ad litteram* 8.1.4 (CSEL 28:1, 231).

and in another way."[78] Those who, like the Manichaeans, unreasonably reject a literal interpretation of Scripture because they are "prompted by an obstinate or just stupid turn of mind" deserve a harsher rebuke. He makes clear, however, that this is a different category of error than the typological view: "But these people of ours," he continues, "who have faith in these divine books, are not prepared to have Paradise understood according to the proper literal sense."[79]

This comparison to the Manichaeans allows Augustine's view on the historicity of Genesis 2–3 to be set in context. He can disagree with a strictly typological view of Eden without considering it to be a rejection of biblical authority, as is Manichaeism; he also does not categorize it as falling outside the boundaries of orthodoxy, since those who hold it "accept the authority of these sacred writings" and "have faith in these divine books." Similarly, when he faces objections from philosophical adversaries in *The City of God* that a physical environment cannot contain earthly bodies, his language is understandably sharper.[80] In other words, Augustine makes a distinction between those who adopt a symbolical/figurative view because they are opposed to the text, and those who take this view sincerely. Origenism and Manichaeism are different animals.

An even more noteworthy concession follows when Augustine wonders how those who reject a geographical Eden can affirm a historical Adam.[81] If we think of Adam as merely a type, Augustine wonders, how far will this go? Must we then regard Seth and Cain and Abel as only figurative?[82] Augustine asserts that we work hard at seeking the literal sense of these chapters. Nonetheless, even here he is unwilling to draw an absolute boundary line: "Certainly, if the bodily things mentioned here could not in any way at all be taken in

[78] *De Genesi ad litteram* 8.1.4 (CSEL 28:1, 231).
[79] *De Genesi ad litteram* 8.1.4 (CSEL 28:1, 231).
[80] *De civitate Dei* 13.18 (CSEL 40.1, 639-41).
[81] *De Genesi ad litteram* 8.1.4 (CSEL 28:1, 231).
[82] *De Genesi ad litteram* 8.1.4 (CSEL 28:1, 231).

a bodily sense that accorded with truth, what other course would we have but to understand them as spoken figuratively, rather than impiously to find fault with holy scripture?"[83] His apologetic concern is evident again here, according to which we must be willing to consider a reinterpretation of Scripture when its veracity is at stake. He continues, however, by warning that if a literal interpretation of Genesis 2–3 lends greater credence to these chapters, we must not be unduly skeptical of it.[84] And overall, the greater portion of Augustine's energy here is spent defending historicity, not considering an alternative to it. He does not rule out a figurative reading of Genesis 2–3, but he is very wary about it.

Later in his literal commentary, Augustine makes a similar concession with regard to the tree of the knowledge of good and evil. In the context of defending the historicity of various elements of the story in Genesis 3, he pauses to note his awareness of a particular alternative view in which Adam and Eve's sin was eating of the tree too early, before they were yet ready for it. In the context of considering this view, Augustine acknowledges, "If by any chance these people mean to take that tree in a figurative sense, and not as a real one with real apples, this opinion may possibly lead to ideas that are agreeable to right faith and the truth."[85] That Augustine can make such a concession reflects once again the strength of his apologetic instinct, along with the extent to which he recognizes a symbolical and literary thrust to the kind of historical narration employed in Genesis 2–3.

Many modern interpreters will be surprised at the flexibility Augustine allows in his understanding of Adam and Eve, particularly in light of his premodern context and his role in shaping the Western tradition of thought on original sin. But we should not so seize on his highly quotable statements to this effect that we mute

[83]*De Genesi ad litteram* 8.1.4 (CSEL 28:1, 232).
[84]*De Genesi ad litteram* 8.1.4 (CSEL 28:1, 232).
[85]*De Genesi ad litteram* 11.41.56 (CSEL 28:1, 376).

his corresponding concerns in the other direction. For all his concessions and openhandedness, Augustine remains unafraid to rebuke over-quick abandonments of historicity in the text. If Augustine's appreciation of the symbolical elements in the text can nuance some conservative readings today, his reluctance to leave off a literal reading should caution revisionist ones.

AUGUSTINE ON THE HISTORY OF THE EVENTS OF GENESIS 2–3

Augustine offers further reasons for holding together both historical reference as well as symbolic import in Genesis 2–3. For instance, in reflecting on the meaning of the trees in the garden, he warns against the danger of too quickly bypassing historical signification for some deeper figurative meaning.[86] It is also important to him to take the four rivers mentioned in Genesis 2:10-14 as real rivers, as opposed to the view he had advocated in his allegorical commentary. At the same time, he is willing to read them like a parable "if any necessity obliged us to take the other things that are told about Paradise only in a figurative and not a proper sense."[87] We should recall that for Augustine, "proper" (*proprie*) in this context is not an evaluative term but a descriptive one, associated with a "literal" interpretation. Moreover, as we saw in chapter 3, reading Genesis 2–3 literally means for Augustine interpreting these chapters as referring to historical events, not as taking all the images and language in a literalistic way.

Augustine will often appeal to angels to protect against over-literalism. For instance, he emphasizes that we cannot understand how God walked in the garden in Genesis 3:8 or spoke to Adam in 2:16-17, but that "nobody who has the least sense of the Catholic faith" will have any doubt that God did these things through an angel.[88] Similarly, he appeals to

[86]E.g., *De Genesi ad litteram* 8.4.8-5.11 (CSEL 28:1, 235-39).
[87]*De Genesi ad litteram* 8.7.13 (CSEL 28:1, 241).
[88]*De Genesi ad litteram* 8.27.50 (CSEL 28:1, 266-67).

angelic agency as an explanation of how the animals came to Adam in Genesis 2:19, "to save us from being crassly literal-minded about how God brought them to Adam."[89]

When Augustine gets to Genesis 3, he feels it necessary to reiterate his guiding principle that we should interpret this story literally when possible, but figuratively when it cannot be understood literally without absurdity.[90] He will retain this balance throughout his treatment of this chapter, affirming an essential historicity while allowing some flexibility in the assessment of which elements are figurative. He thinks the devil used an ordinary serpent to tempt Adam and Eve: "It was the devil himself who spoke in the serpent, using it like an organ, and moving its nature to give expression to the verbal sounds and bodily gestures by which the woman would be made to understand what he wished to persuade her to do."[91] The serpent itself was unaware of what it was saying;[92] and the reference to the serpent as the "most cunning of all the wild beasts" (Gen. 3:1) applies to the devil, not the animal.[93] At the same time, he emphasizes that it was God's choice to allow a serpent for this end rather than the devil's;[94] and he stipulates a "deeper meaning" to this element in this story, calling it a "prophecy."[95]

A similar evenhandedness marks his subsequent exposition. He interprets the phrase "your eyes shall be opened" in Genesis 3:5 (and 3:7) metaphorically, but insists that there is no reason "why the expression should lead us to treat the whole story as allegorical."[96] He insists that God did curse the serpent, but regarding the words spoken in this curse allows that "it is left to the reader's free judgment whether

[89]*De Genesi ad litteram* 9.14.24 (CSEL 28:1, 285).
[90]*De Genesi ad litteram* 11.2.1 (CSEL 28:1, 334-35).
[91]*De Genesi ad litteram* 11.27.34 (CSEL 28:1, 360).
[92]*De Genesi ad litteram* 11.28.35 (CSEL 28:1, 360-61).
[93]*De Genesi ad litteram* 11.29.36 (CSEL 28:1, 361).
[94]*De Genesi ad litteram* 11.3.5 (CSEL 28:1, 337).
[95]*De Genesi ad litteram* 11.12.16 (CSEL 28:1, 344-45).
[96]*De Genesi ad litteram* 11.31.41 (CSEL 28:1, 364).

they should be taken literally or figuratively."[97] As he proceeds, Augustine ultimately inclines toward a figurative interpretation of this curse, comparing this method of interpretation to the one that he attempted in his allegorical work against the Manichaeans.[98] He also affirms a figurative interpretation of the curse on the woman.[99] The curse on the man, he maintains, has both a historical and prophetic significance. We should therefore not hesitate to take these words "first and foremost in their proper historical sense," while at the same time "a prophetic signification is to be looked for and expected, and it is this that the divine speaker here has chiefly in mind."[100] Regarding God providing tunics for Adam and Eve after their fall, Augustine says we may understand the words spoken in this context either metaphorically or literally, but we must not reduce "its actually being said" to a metaphor.[101] With regard to the angel with the flaming sword in Genesis 3:24, Augustine believes this was a real heavenly power in the visible paradise, even if it may also signify something about spiritual paradise.[102]

Taken as a whole, Augustine's position on the various details of Genesis 2–3 reflects considerable balance: he insists on the story's essential historicity, but is sensitive to the danger of taking the symbol-laden and/or stylized language employed by the text too literally. To lose historicity is to give up what must be defended; to interpret with literary insensitivity is to make our defense more vulnerable to skeptical assault. In addition, this second mistake often assumes that the Bible is speaking more comprehensively than it intends, another danger to which Augustine is sensitive. In *The City of God*, for instance, after noting that some wonder why Cain built a city when there were only three people on earth, Augustine responds:

---

[97]*De Genesi ad litteram* 11.36.49 (CSEL 28:1, 371).
[98]*De Genesi ad litteram* 11.36.49 (CSEL 28:1, 371).
[99]*De Genesi ad litteram* 11.37.50 (CSEL 28:1, 372).
[100]*De Genesi ad litteram* 11.38.51 (CSEL 28:1, 373).
[101]*De Genesi ad litteram* 11.39.52 (CSEL 28:1, 374).
[102]*De Genesi ad litteram* 11.40.55 (CSEL 28:1, 375).

"The writer of the sacred history does not necessarily mention all the men who might be alive at that time, but those only whom the scope of his work required him to name."[103] He proceeds to emphasize that the early chapters of Genesis are concerned primarily with the lineage leading to Abraham. Here again we see Augustine's dual concerns with respect to the early chapters of Genesis: on the one hand, he wants to maintain their historicity; on the other, he wants to interpret them with literary and apologetic sensitivity.

AUGUSTINE ON THE NATURE OF ADAM AND EVE

Augustine is often contrasted with Irenaeus as having a higher view of the perfection of Adam and Eve, and the pre-fallen world more broadly, as opposed to Irenaeus's "developmental" account.[104] The differences between them, however, are easy to exaggerate. Along with many other exegetes in the early church, Augustine held that pre-fallen Adam and Eve possessed a *conditional immortality*—that is, he thinks that if they had obeyed God they would have been translated into an immortal existence. As he explains in *The City of God*, "If they discharged the obligations of obedience, an angelic immortality and a blessed eternity might ensue, without the intervention of death."[105] In his literal commentary on Genesis, he describes this as a "wonderful condition" in which "it would have been impossible for them to be tried by disease or altered by age," and claims that it "was to be bestowed upon them through the mystical virtue in the tree of life."[106] Augustine evidently does not believe that Adam and Eve were immortal apart from the transformation that would have occurred through the tree of life. At the same time, he still regards the conditions of their existence in this

[103]*De civitate Dei* 15.8 (CSEL 40.2, 72).

[104]A more careful account of Irenaeus's view of the fall, however, with intriguing implications for current discussions, is offered in A. N. S. Lane, "Irenaeus on the Fall and Original Sin," in *Darwin, Creation and the Fall: Theological Challenges*, ed. R. J. Berry and T. A. Noble (Nottingham: Apollos, 2009), 140-48.

[105]*De civitate Dei* 13.1 (CSEL 40.1, 615).

[106]*De Genesi ad litteram* 11.32.42 (CSEL 28:1, 365-66).

probationary state as significantly different from current human experience. For example, he ponders why Adam was given food to eat along with the other animals, since "surely before sin he would not have needed such nourishment."[107] He conceives of pre-fallen human life to have also lacked the animal instinct of sexual desire, and speculates that Adam and Eve could have populated the world apart from sex.[108]

Nonetheless, it is clear that Augustine did not regard pre-fallen Adam and Eve to be perfect and in need of no further development. In the *Enchiridion* he claims that God placed Adam "in the bliss of paradise as if in the shadow of life, from which he was to rise to better things if he had preserved his state of justice."[109] The language of "the shadow of life" suggests that Augustine conceives of the paradise of Eden not as an end in itself, but as a step toward a greater end—perhaps a kind of representation or type of the immortality that Adam and Eve would have attained had they not fallen. Later, he stipulates that the "better things" to which Adam would have arisen include the state of being unable to sin. He further suggests that this transition would not have occurred immediately upon their obedience, but "at the appropriate time after he had had children . . . without the intervention of death."[110] In his literal commentary as well, Augustine argues that if Adam and Eve had not sinned, their "ensouled animal bodies" would have been translated into spiritual bodies without the intervention of death.[111] He draws attention to the examples of Enoch and Elijah in order to envision how Adam and Eve might have begotten children who would populate the world before being translated into God's presence apart from physical death.[112] Thus, Augustine apparently did not think of the pre-fallen state of humanity

[107]*De Genesi ad litteram* 3.21.33 (CSEL 28:1, 88).
[108]*De Genesi ad litteram* 3.21.33 (CSEL 28:1, 88).
[109]*Enchiridion* 8.25, in *On Christian Belief*, 289.
[110]*Enchiridion* 28.104, in *On Christian Belief*, 334.
[111]*De Genesi ad litteram* 9.3.6 (CSEL 28:1, 272).
[112]*De Genesi ad litteram* 9.6.10-11 (CSEL 28:1, 273-75).

as permanent—rather, he seems to have thought of the existence of Adam and Eve in Eden as a kind of provisional testing period.

Nor does Augustine think Adam and Eve were fully mature in Eden. Elsewhere, for instance, he stipulates that Adam was created in a state of innocence, but not of wisdom. In *On Free Will*, he addresses the objection, "If the first man was created wise, then why did he allow himself to be seduced?"[113] By way of response, Augustine identifies an "intermediate condition" between wisdom and folly, comparable to human infancy: "Just as it is senseless to call an infant stupid, so it would be even more absurd to call it wise, even though it is already a human being. It clearly follows from this that the nature of man allows for some intermediate condition, which cannot rightly be called either wisdom or folly."[114] Augustine compares the danger of associating this intermediate position with folly to the error of associating a newborn animal that cannot yet see with a blind adult animal.[115] Thus, for Augustine, it is a grave error to cast any fault on the nature of Adam and Eve, as though this intermediate condition were flawed in any way. But it is also wrong to conceive of pre-fallen Adam and Eve as perfect and fully developed. They were not fools, but neither were they yet wise.

ADAM AND EVE TODAY

How might we leverage Augustine's views of Genesis 2–3, and his conception of *rationes seminales*, to help current evangelical perspectives of Adam and Eve? As we have noted, this is a currently controversial area in evangelical thought. We can construct a taxonomy of three broad instincts on the question:

1. Because we must accept evolution, we must reject belief in a historical Adam and Eve (*evolution, therefore no Adam*).

---

[113]*De libero arbitrio* 3.24 (CSEL 74, 148).
[114]*De libero arbitrio* 3.24 (CSEL 74, 148).
[115]*De libero arbitrio* 3.24 (CSEL 74, 148-49).

2. Because we must maintain a historical Adam and Eve, we must reject evolution (*Adam, therefore no evolution*).
3. We may maintain belief in a historical Adam and Eve while considering harmonization with some forms of evolution (*Adam and evolution*).

As the temperature of the discussion has risen, views on Adam and Eve in relation to evolution seem to be growing more polarized. Thus, much of the current evangelical discussion is marked by warfare between instincts one and two, whereas in previous generations, even in many relatively conservative circles, instinct three was more commonly considered. We will explore this third instinct more fully in a moment. But first, we must draw Augustine into critical interaction with the first two instincts as they are represented in the current evangelical scene.

Obviously, Augustine cannot settle these issues in any decisive way. He long predates them. Nonetheless, his voice may provide a helpful contribution to the discussion, particularly to the extent that we are willing to consider the overall posture of his treatment of Adam and Eve, and the range of concerns that are reflected in it. So here we will suggest that the principles of Augustine's thought surveyed throughout this chapter would discourage us from accepting instinct one (*evolution, therefore no Adam*) but would leave us open both to instinct two (*Adam, therefore no evolution*) and instinct three (*Adam and evolution*), to follow the evidence wherever it directs us between these options.

INSTINCT ONE: EVOLUTION, THEREFORE NO ADAM

This position is familiar as it advanced in secular contexts, where harmonization efforts between Adam and Eve and evolutionary theory are often mocked as "ludicrous and tortuous"[116] because of "the clear results of populations genetics that [Adam and Eve] could *not* have

[116]Jerry Coyne, "Adam and Eve: Theologians Squirm and Sputter," https://whyevolutionistrue.wordpress.com/2011/09/08/adam-and-eve-theologians-squirm-and-sputter. Cf. Jerry Coyne, *Why Evolution is True* (New York: Viking, 2009).

existed."[117] But *evolution, therefore no Adam* is also representative of an increasing number of evangelicals. To be clear, though, this is not the position of all who go under the label evolutionary creationist/ theistic evolutionist. For instance, in the contemporary ministry BioLogos, there are a diversity of views about Adam and Eve, and under their umbrella are included a number of those who affirm a historical Adam and Eve and a historical fall.[118]

However, some of those voices advocating for an evolutionary account of human origins take this to require the rejection of a historical Adam and Eve. In his contribution to the recent influential book *Adam and the Genome*, Scot McKnight argues that a historical Adam, at least in the sense understood by many Christians today, is actually a recent concept. Throughout *Adam and the Genome*, McKnight employs a taxonomy of titles for Adam based on his different functions (literary Adam, genealogical Adam, archetypal Adam, and so forth).[119] He draws a number of useful insights about how Adam and Eve were understood in their ancient Near Eastern context and in subsequent Jewish texts. Nonetheless, although he states early on that "literary" does not mean either "fictional" or "historical,"[120] in his conclusions he often appears to set a literary Adam or archetypal Adam at odds with a historical Adam. After developing his final thesis, "the Adam of Paul was not the historical Adam,"[121] McKnight concludes the book by urging that the very notion of a historical Adam and Eve didn't even exist in Paul's day:

---

[117]Jerry Coyne, "Pete Enns, BioLogos, and Adam and Eve: Why Accommodationism Won't Work," https://whyevolutionistrue.wordpress.com/2012/01/26/pete-enns-biologos-and-adam-and-eve-why-accommodationism-wont-work.

[118]See Kenneth Keathley, J. B. Stump, and Joe Aguirre, eds., *Old-Earth or Evolutionary Creation? Discussing Origins with Reasons to Believe and BioLogos*, BioLogos Books on Science and Christianity (Downers Grove, IL: IVP Academic, 2017), 49-67; and "Were Adam and Eve Historical Figures?," https://biologos.org/common-questions/human-origins/were-adam-and-eve-historical-figures, accessed April 11, 2018.

[119]He defines some of these labels on page 108, and deepens his definition of "literary" on 210n37.

[120]Venema and McKnight, *Adam and the Genome*, 118.

[121]Venema and McKnight, *Adam and the Genome*, 188.

> We need to give far more attention than we have in the past to the various sorts of Adams and Eves the Jewish world knew. One sort that Paul didn't know because it had not yet been created was what is known today as the *historical* Adam and Eve. Literary Adam and Eve, he knew; genealogical Adam and Eve, he knew; moral, exemplary, archetypal Adam and Eve, he knew. But the *historical* Adam and Eve came into the world well after Paul himself had gone to his eternal reward, where he would have come to know them as they really are.[122]

But there is a danger here of an unnecessary dichotomy between Adam's historicity, on the one hand, and his theological/literary deployment, on the other.[123] It is specious to assume that a literary figure must be *only* literary: a literary or archetypal Adam may also be a historical Adam, or may not; that must be settled on other grounds. By comparison, to borrow Joseph Fitzmyer's classic categories, the theological usage of Jonah in Matthew 12:40 or of Melchizedek in Hebrews 7 does not require Jonah and Melchizedek to be ahistorical.

Part of the difficulty here, to be fair, is that McKnight is operating with a rather heavily loaded meaning of the term *historical Adam*, involving seven distinct assertions.[124] For instance, included within the purview of the label *historical Adam* is not only the biological and genetic role of Adam and Eve as the first two human beings who pass on a sin nature to all of us, but the *consequence* of these claims: "If one denies the *historical* Adam, one denies the gospel of salvation."[125] But there are many, of course, who affirm a historical Adam without

---

[122]Venema and McKnight, *Adam and the Genome*, 191, italics his.

[123]I do appreciate *Adam and the Genome*'s effort to engage the genetic science, a task that some evangelicals simply ignore. While I found the book more helpful for learning about the science of genetics than for accounting for this data theologically, I share the authors' sensitivity to the crisis of young people losing their faith over evolution. This is indeed a real problem, and the book repeatedly draws attention to it—both the afterword by Daniel Harrell and the foreword by Tremper Longman touch on it, and it comes up regularly throughout (Alexis de Tocqueville's description of his own struggle with doubt, recounted by McKnight on page 103, provides one poignant portrait).

[124]Venema and McKnight, *Adam and the Genome*, 107-8, 188-89.

[125]Venema and McKnight, *Adam and the Genome*, 188.

necessarily affirming all these claims *about* the historical Adam. One is left wondering what McKnight would conclude about other, more restricted (and arguably more judicious) affirmations of a "historical Adam."

McKnight's engagement with the Jewish literature may also downplay the extent to which there is potential overlap between contemporary affirmations of a historical Adam and the role that Adam plays in historical texts. For instance, although Sirach 25:24 (NRSV) claims that "from a woman [Eve] sin had its beginning, and because of her we all die," McKnight claims that the Adam of Sirach is archetypal and moral, and "while there are traces of a genealogical Adam, the Adam of Sirach is not the historical Adam who plays such an important role in Christian theology."[126] Likewise, McKnight acknowledges that Wisdom of Solomon references "the first-formed father of the world, when he alone had been created" (Wis 10:1 NRSV) and claims that each human is "a descendant of the first-formed child of earth" (Wis 7:1 NRSV). But he insists that the author "is not thinking scientifically, biologically, genetically, or even historically but instead is affirming his tradition and theologizing in light of it."[127] McKnight admits that Philo probably believed that Adam was the first human, since he repeatedly calls Adam "the first man" and "ancestor of our race" throughout *On the Creation*. But he downplays this conclusion, emphasizing Philo's strange beliefs and other uses of Adam, and leaving the historical Adam out of his summary of Philo.[128] In these various texts, it seems eminently possible that Adam is playing a theological and/or literary role while still being conceived of as a historical person.

It is therefore difficult to agree with McKnight's conclusion that "no author cared about a 'historical' reading; each author adapted and

[126]Venema and McKnight, *Adam and the Genome*, 156.
[127]Venema and McKnight, *Adam and the Genome*, 158.
[128]See Venema and McKnight, *Adam and the Genome*, 160-61.

adopted and adjusted the Adam of Genesis;"[129] and that "the historical Adam that Christians now believe in has yet to make his appearance on the pages of history. . . . The construct Christians use when they speak of the historical Adam is not to be found in the Old Testament or in other Jewish sources."[130] This claim can only be made if we isolate certain ways of thinking about a historical Adam today, and then set this way of thinking about Adam at odds with other ways of thinking about Adam that are potentially complementary to it. The fact that some Jewish texts theologize differently about Adam from some contemporary evangelicals and fundamentalists doesn't mean that *all* concern about Adam's historicity is novel.

The concern of an unnecessary dichotomy between a historical Adam and the theological and literary deployment of Adam also arises in the claims of Peter Enns. Enns urges understanding the Adam of Genesis 2–3 in light of Israelite national identity.[131] The concern of these chapters, he suggests, has less to do with our modern interest in biological origins and more to do with Israel making theological sense out of the Babylonian captivity: "The primary question Israel was asking was not, 'Where do we come from?' (a scientific curiosity), but 'Where do we come from?' (a matter of national identity)."[132] Thus, Genesis 2–3 portrays the nature of wisdom, and confronts every reader with a choice: "The story of Adam becomes a story for 'every Israelite,' those who are daily in a position of having to choose which path to take: the path of wisdom or the path of foolishness."[133] Therefore, as Enns colorfully puts it, "If evolution is right about how humans came to be, then the biblical story of Adam and Eve isn't. . . . Dragging the Adam and Eve story into the evolution discussion is as misguided as using the stories of Israel's monarchy to

[129] Venema and McKnight, *Adam and the Genome*, 168.
[130] Venema and McKnight, *Adam and the Genome*, 169.
[131] Peter Enns, *The Evolution of Adam: What the Bible Does and Doesn't Say About Human Origins* (Grand Rapids: Brazos, 2012), 140-42.
[132] Enns, *The Evolution of Adam*, 142.
[133] Enns, *The Evolution of Adam*, 142.

rank the Republican presidential nominees."[134] But as with McKnight, this seems to unnecessarily dichotomize the historical and typological. As Matthew Levering points out, even if Genesis is primarily concerned with Israel, it does not follow that it therefore has *no* interest in the origins of human sin and death.[135] Appreciating Enns's emphasis on an Adam-Israel typology, we may still ask: Why can the story of Genesis 2–3 not also have a larger reference? Is interest in human origins an *exclusively* modern interest?

Augustine can help us at this point. We have already seen his concern throughout his literal commentary to take the historicity of Genesis 2–3 seriously while also appreciating the text's symbolical and literary features. He adopts a similar posture in his other writings. For instance, to use one example in *The City of God*, Augustine reiterates his acceptance of the "spiritual sense" of the various elements and events associated with the garden of Eden. But he then follows this discussion with a strong warning against pitting the spiritual and historical against one another, drawing a comparison with other historical events that take on a figurative meaning in the Old Testament:

> As if there could not be a real Paradise! As if there never existed these two women, Sarah and Hagar, nor the two sons who were born to Abraham, the one of the bond woman, the other of the free, because the apostle says that in them the two covenants were prefigured; or as if water never flowed from the rock when Moses struck it, because therein Christ can be in a figure, as the same apostle says, "Now that rock was Christ!"[136]

Augustine maintains similar views of Noah and the ark, insisting on both their symbolical and historical meanings.[137] Now, again, Augustine did not face the precise issues we are facing today, and so it

[134]Peter Enns, "Once More, With Feeling: Adam, Evolution and Evangelicals," www.huffingtonpost.com/pete-enns/adam-evolution-and-evangelicals_b_1219124.html; cf. also Enns, *The Evolution of Adam*.

[135]Matthew Levering, *Engaging the Doctrine of Creation: Cosmos, Creatures, and the Wise and Good Creator* (Grand Rapids: Baker Academic, 2017), 248-49.

[136]*De civitate Dei* 13.21 (CSEL 40.1, 645).

[137]*De civitate Dei* 15.27 (CSEL 40.2, 118-22).

would be unwise to put too much weight on his words here, as though they simply settled the issue. Nonetheless, Augustine may at least deepen our appreciation for the possibility of holding together symbolical/figurative and historical meanings of various biblical motifs and types. With Augustine's exegesis in mind, it is more difficult to assume that the rich array of theological and literary roles that Adam and Eve have come to play in subsequent texts should somehow amount to a strike against their historicity.

As we have observed, Augustine is aware of the possibility of a symbolical or figurative Adam—an Adam who is *only* a type, not an actual person in history. (Recall the first several pages of book 8 of the literal commentary, which should restrain McKnight's claims that concerns about Adam's historicity are recent.) Moreover, while Augustine does not cast this error with the Manichaeans, his opposition to it is vigorous. He observes that Genesis 2–3 is like 1 and 2 Kings, not like Song of Solomon, in that it "quite simply tells of things that happened."[138] He chides those who point to the supernatural undercurrent in Genesis 2–3 to claim that history begins in Genesis 4, pointing out that such miraculous events continue on after these chapters—"as though forsooth we are quite familiar with people living as many years as they did, or with things like Enoch being taken, or a very old and barren woman giving birth, and other things of that sort!"[139] Augustine points to Adam's activities in Genesis, including his fathering of children, to urge his readers not to follow those who regard him as only figurative: "If he too is to be understood figuratively, who was it who begot Cain and Abel and Seth? Or did they also exist in figure, not also as real human beings?"[140]

In light of these difficulties, Augustine urges that those persuaded of a figurative Adam should "try hard with us to take all these primordial

---

[138]*De Genesi ad litteram* 8.1.2 (CSEL 28:1, 229).
[139]*De Genesi ad litteram* 8.1.2 (CSEL 28:1, 230).
[140]*De Genesi ad litteram* 8.1.4 (CSEL 28:1, 231).

events of the narrative as actually having happened."[141] Augustine offers in return that, when they have done so, we would then "support them as they turned their minds next to working out what lessons these things have for us in their figurative meaning, whether about spiritual natures and experiences or even about events to come in the future."[142] Notice again here Augustine's dexterity in reaching out for both historical and figurative meanings simultaneously.

Another problem with removing the historicity of the fall is the challenge it poses for accounting for original sin and the human condition, a topic enormously important to Augustine's theology. It is one thing to posit the need for development in the understanding of original sin—as we will explore below in discussing its mechanism of transmission. But to do away with a historical fall altogether runs the risk of blaming God for the human condition by removing the human agency that produced original sin. This is the danger that James K. A. Smith calls "'naturalizing' sin."[143] Similarly, Henri Blocher warns that evil must always be seen as an "alien intruder" rather than a part of God's original creation.[144] Denying the historical fall raises this theodicy challenge. That human life is irreparably broken is a basic fact of life that even many secular people will not deny.[145] How, apart from a historical fall, do we avoid placing the blame for this fact at God's feet?

A common feature of evolutionary creationist argumentation is that the science favoring polygenism is as strong as that favoring an

[141]*De Genesi ad litteram* 8.1.4 (CSEL 28:1, 231-32).

[142]*De Genesi ad litteram* 8.1.4 (CSEL 28:1, 232).

[143]James K. A. Smith, "What Stands on the Fall? A Philosophical Exploration," in *Evolution and the Fall*, ed. William T. Cavanaugh and James K. A. Smith (Grand Rapids: Eerdmans, 2017), 63; as Smith quips later on the same page: "Making sin original is *not* the doctrine of original sin" (italics his).

[144]Blocher, "The Theology of the Fall and the Origins of Evil," in *Darwin, Creation and the Fall*, 160-64.

[145]Henri Blocher, *Original Sin: Illuminating the Riddle*, New Studies in Biblical Theology 5 (Downers Grove, IL: InterVarsity Press, 1997), 20: "Even those who oppose the church dogma of original sin concur in this basic assessment of our reality. It would be hard to close one's eyes to the data of experience." Later (83-103), Blocher insightfully considers how original sin is a riddle that nonetheless makes sense of our experience as human beings.

ancient universe, and thus we should be open to evolution for the same reasons that we should be open to the universe's age.[146] This is a fair and consistent appeal, and theologians should not respond to the claims of evolutionary science with less patience or scrutiny than they respond to other scientific claims. At the same time, some of those urging revision to traditional views of Adam and Eve seem to understate the theological implications of these revisions, and at times it seems that theological adjustments to the science are urged more eagerly than potential scientific adjustments in light of theology.[147] While we should listen carefully to scientific claims, Augustine would discourage us from adopting any posture in which the science is playing offense and the Bible is playing defense. Theology can speak to science, as well as listen.

INSTINCT TWO: ADAM, THEREFORE NO EVOLUTION

What about the second view, which, as we have seen, is widely spread throughout evangelicalism?[148] Before inquiring into how Augustine would respond to this view, I should clarify that my goal in this chapter is not to use Augustine to simply score a point for the evolutionary creationism camp. As we have seen, Augustine's views do not appear neatly serviceable to *any* contemporary creation view. Moreover, I am personally uncertain as to the exact extent of the explanatory reach of evolutionary mechanisms, and I consider it a question outside my training and calling. After much agonizing study over this issue, however, I have come to the view that the church needs to evaluate evolution with greater humility and carefulness

[146]For example, Jim Stump makes this point eloquently in his contribution to *Sapientia*'s book symposium on Ted Cabal's *Controversy of the Ages*: "Controversy and Conversation," http://henrycenter.tiu.edu/2017/12/controversy-and-conversation.

[147]As we have noted, *Adam and the Genome* helpfully draws attention to the importance of interpreting Scripture with honesty, respect, and sensitivity to the student of science, while its treatment of Scripture's primacy (and we would add the word *authority*) seems underemphasized (93-109).

[148]For a good representation of this position, see William VanDoodewaard, *The Quest for the Historical Adam: Genesis, Hermeneutics, and Human Origins* (Grand Rapids: Reformation Heritage, 2015), 281-312.

than we have sometimes exhibited, and I do think that Augustine can be a helpful stimulus to this end. Thus, here I would suggest, not so much that Augustine would oppose the *Adam, therefore no evolution* view, but that he would offer caution to the extent that it is put forward as the only possible option. In other words, I think Augustine would encourage us to be careful in adjudicating between the second and third instincts identified above.

Such openness is not often taken to be Augustine's legacy, to put it mildly. For those who regard traditional notions of original sin and Adam and Eve as a problem awaiting revision, no one is blamed more than Augustine.[149] In the other direction, defenses of original sin often take Augustine for their patron saint. But Augustine's legacy is more complicated than is sometimes appreciated in polemical contexts. As we saw in chapter 2, Augustine advocated for a careful and humble engagement with scientific claims, and was willing to reconsider theological interpretations that were in conflict with such claims, particularly when the science in question was well established. Moreover, while Augustine maintained the historicity of Genesis 2–3, he recognized the stylized nature of the text, and he exhibited remarkable flexibility in interpreting its various details.

Take, as an example, the issue of the "dirt" or "dust" of Genesis 2:7: "The LORD God formed the man of dust from the ground." In evangelical circles, it is often asserted or assumed that this passage has in view literal dust—the tiny particles that we sweep up with brooms and that collect on the soles of our shoes. Augustine, on the other hand, though holding to a de novo creation of Adam, worries about understanding this too literally. To summarize from an earlier discussion, Augustine emphasizes the diversity of the meanings of this term; he works hard to avoid a simplistic understanding of divine agency in

---

[149]E.g., Patricia A. Williams, *Doing Without Adam and Eve: Sociobiology and Original Sin* (Minneapolis: Fortress, 2001), 40-47; more carefully and more critically, Tatha Wiley, *Original Sin: Origins, Developments, Contemporary Meanings* (New York: Paulist, 2002), 56-75.

Adam's creation (an error he detects among the Manichaeans); he is open to the possibility that Adam was created as an infant, as subsequent human beings are; he is comfortable with interpreting various other details of the garden of Eden allegorically, such as the trees and fruits and rivers; and he emphasizes the creation of Eve from Adam's rib as a prophetic signification of sacrament and marriage (even while insisting that it can be historical as well).

Perhaps most relevant to the contemporary discussion is Augustine's awareness of how other Christians (like Origen) have understood Adam and Eden figuratively, and his recognition of this as a possible interpretation within the church. As we observed earlier, Augustine distinguishes between those like the Manichaeans who reject a literal interpretation because they are obstinate and opposed to the faith, and "these people of ours who have faith in these divine books, and are not prepared to have Paradise understood according to the proper literal sense."[150] Augustine ultimately defends the historicity of Genesis 2–3, but he also appears to regard a figurative interpretation of Adam and Eden as possible among faithful Christians. Even more striking, as we saw earlier, is Augustine's willingness to qualify his defense of a historical Adam and physical Eden by envisioning the possibility that he is wrong: "Certainly, if the bodily things mentioned here could not in any way at all be taken in a bodily sense that accorded with truth, what other course would we have but to understand them as spoken figuratively, rather than impiously to find fault with holy scripture?"[151] We have also noted his willingness to make the same claim about particular details in Genesis 2–3, such as the tree of the knowledge of good and evil: "If by any chance these people mean to take that tree in a figurative sense, and not as a real one with real apples, this opinion may possibly lead to ideas that are agreeable to right faith and the truth."[152]

---

[150]*De Genesi ad litteram* 8.1.4 (CSEL 28:1, 231).
[151]*De Genesi ad litteram* 8.1.4 (CSEL 28:1, 232).
[152]*De Genesi ad litteram* 11.41.56 (CSEL 28:1, 376).

Augustine's openness on the creation of Adam will likely be surprising to those who take Genesis 2:7 as a photographic depiction of a de novo creation of Adam. It may help us appreciate his caution, however, to consider how the word *dust* (Hebrew *aphar*) in this verse continues to be used in the rest of Scripture. In Genesis 3:19, for instance, Adam is told at the conclusion of his curse, "For you are dust, and to dust you shall return." The use of this term in the present tense for Adam's living body ("you are dust"), as well as for the prospect of death ("to dust you shall return"), raises questions about how exactly this language is being used with reference to Adam's origins. Quite obviously, Adam's living body ("you are dust") cannot be the kind of dust we sweep with a broom. Moreover, we find the same language applied to all human beings elsewhere in Scripture. Thus, Ecclesiastes 3:20 says, "All are from the dust, and to dust all return"; Psalm 103:14 says that God "knows our frame; he remembers that we are dust;" Job 10:9-10 combines the dust imagery with the clay/potter imagery for Job's own creation and death; and Paul picks up on the same imagery for all human beings in 1 Corinthians 15:48, stipulating that all subsequent humans are from the dust because Adam was from the dust (Greek *choikos*): "As was the man of dust, so also are those who are of the dust, and as is the man of heaven, so also are those who are of heaven. Just as we have borne the image of the man of dust, we shall also bear the image of the man of heaven."

All this makes it more difficult to insist that "from the dust" necessarily entails a de novo process.[153] It seems at least possible to take Genesis 2:7 as making a more general claim about the nature of Adam,

[153]For further treatment of the meaning of "dust" in Genesis 2:7, see John Walton, *The Lost World of Adam and Eve: Genesis 2–3 and the Human Origins Debate* (Downers Grove, IL: IVP Academic, 2015), 72-77. See also J. Richard Middleton, "Reading Genesis 3 Attentive to Human Evolution," in *Evolution and the Fall*, 78-80, who treats the meaning of "dust" in connection to a helpful overview of the meaning of "life" and "death" in Genesis 2–3. I recognize that I have focused less on Eve's creation from Adam's rib here; that is addressed more fully in these resources.

akin to the claim being advanced about all human beings in Ecclesiastes 3:20. Perhaps, for instance, the "dust" in Genesis 2:7 is not referring to the physical substance with which he was made, but rather pointing to his physicality *as such*, and consequent mortality. If so, then more caution must be exhibited regarding how we apply this passage to the kinds of questions that come up when discussing human evolution. At the very least, it would seem wise not to elevate the issue of the mechanism for Adam's creation to the same level of importance as his historical existence and role.

Augustine would almost certainly encourage this kind of caution. As we have seen, he resists denying the essential historicity of the narrative of Genesis 2–3 but is more openhanded on the interpretation of various details in the story, including the nature of Adam's creation from the "mud," along with other details that we have focused on less above, such as Eve's creation from his rib. In particular, it is difficult to read his patience for Origen's supposal of a figurative Adam, or his own allowances for a figurative interpretation of such details as the trees in the garden of Eden, and then situate Augustine exclusively within option two (*Adam, therefore no evolution*). After all, the issues that distinguish options two and three are broadly those on which Augustine is more circumspect.

Now, the notion that Augustine would potentially be open to evolution will doubtless be shocking to some. In many contemporary evangelical circles, any form of evolution, whether or not one retains a historical Adam and Eve, is regarded as a clear enemy of the gospel. It may therefore be helpful in the next section to draw Augustine's treatment of Adam and Eve into contact with how other Christians since Darwin's time have struggled with these issues. Many Christians, operating in many different traditions, have found themselves pursuing neither of these first two instincts we have here surveyed. In other words, in the ways that Augustine is surprising to us, he is not alone.

INSTINCT THREE: ADAM AND EVOLUTION

Although the theory of biological evolution was controversial within the Roman Catholic Church in the decades following Darwin, it grew in favor throughout the twentieth century. Since around 1950 it has been broadly accepted by the highest authorities within Catholicism,[154] and Catholic theologians such as Karl Rahner and Kenneth Kemp have been leading voices in the attempt to harmonize Genesis 2–3 with evolutionary science.[155] Evolution has historically enjoyed a broad though not universal support among the Orthodox as well, with Orthodox Christians making important contributions in the development of evolutionary science. Theodosius Dobzhansky, for example, who is Russian Orthodox, was one of the leading figures in twentieth-century evolutionary biology, with his book *Genetics and the Origin of Species* playing a critical early role in the acceptance of the "modern synthesis" of Charles Darwin's notion of natural selection with Gregor Mendel's conception of genetic inheritance.[156] Interestingly, evangelical conversions to Orthodoxy may be complicating the status of evolution in Orthodox circles—for example, Gayle Woloschak observes that there are more Orthodox Christians who accept evolution than who reject it, but worries about a recent "infiltration into the Church of fundamentalist tendencies that have been foreign to Orthodoxy in the past, coming predominantly from Protestant Evangelicals."[157]

But it is not only Catholic and Orthodox Christians who have found ways to accommodate evolutionary science into their theology.

---

[154]For a historical overview of Catholic teaching on evolution, see John P. Slattery, *Faith and Science at Notre Dame: John Zahm, Evolution, and the Catholic Church* (Notre Dame, IN: University of Notre Dame Press, 2019), 5-24.

[155]For a helpful overview of the contributions of Rahner and Kemp, see Levering, *Engaging the Doctrine of Creation*, 235-41.

[156]Theodosius Dobzhansky, *Genetics and the Origin of Species* (New York: Columbia University Press, 1937).

[157]Gayle E. Woloschak, "The Compatibility of the Principles of Biological Evolution with Eastern Orthodoxy," *St. Vladimir's Theological Quarterly* 55.2 (2011): 215-16.

Instinct three has been well represented among Protestants as well since the days of Darwin, including conservative and evangelical Protestants. R. J. Berry and T. A. Noble observe that opposition to evolution "was not the stance of Evangelical theologians such as James Orr or B. B. Warfield in the decades after Darwin; it was the position adopted by grassroots fundamentalists in the 1920s in reaction to the way in which T. H. Huxley and others had turned Darwin's science into propaganda."[158] Similarly, Tim Keller notes:

> Despite widespread impression to the contrary, both inside and outside the church, modern Creation Science was not the traditional response of conservative and evangelical Protestants in the nineteenth century when Darwin's theory first became known. . . . R. A. Torrey, the fundamentalist editor of *The Fundamentals* (published from 1910–1915, which gave definition to the term "fundamentalist"), said that it was possible "to believe thoroughly in the infallibility of the Bible and still be an evolutionist of a certain type." . . . The man who defined the doctrine of Biblical inerrancy, B. B. Warfield of Princeton (d. 1921) believed that God may have used something like evolution to bring about life-forms.[159]

Evaluation of evolutionary science in the church continues to play out differently in different parts of the world today. In other words, there are not only generational differences to how this debate is approached, but geographical differences. N. T. Wright, for instance, draws attention to the sociopolitical undercurrents of this discussion as they play out differently in American and British evangelicalism.[160] He concludes his essay humorously to make the point:

---

[158]R. J. Berry and T. A. Noble, "Foreword," in *Darwin, Creation, and the Fall*, 11. Cf. Ronald L. Numbers, *The Creationists: From Scientific Creationism to Intelligent Design* (1992; expanded ed., Cambridge, MA: Harvard University Press, 2006), 7: "By the late nineteenth century even the most conservative Christian apologists readily conceded that the Bible allowed for an ancient earth and pre-Edenic life."

[159]Tim Keller, *The Reason for God: Belief in an Age of Skepticism* (New York: Dutton, 2008), 262.

[160]N. T. Wright, "A British Reflection on the Evolution Controversy in America," in *How I Changed My Mind About Evolution*, 131-37.

> There is an old story about a gang of youths in Belfast stopping an Indian gentleman on the street. "Are you a Catholic or a Protestant?" they demand. "I'm a Hindu!" he answers. "Okay," they say, "but are you a Catholic Hindu or a Protestant Hindu?" So, am I a fundamentalist creationist or an atheistic scientist? Answer: I'm a Brit.[161]

In general, the third instinct mentioned above—*Adam and evolution*—has a wider attestation throughout the global church since the time of Darwin than is often realized. As J. I. Packer summarizes the scene:

> A spectrum of views exists [on the relation of creation and evolution]. At one end are scientists who believe in evolution, but not in God. . . . At the other end of the spectrum are believers in God who do not believe in evolution; these vary among themselves in the way they conceive evolution and understand the biblical witness to God's work of creation. Between the two extremes are many, professional scientists and theologians as well as men and women in the street, who believe in both God and evolution, seeing evolution as one element in God's way of making and ordering his world.[162]

Such harmonization efforts were common among initial reactions to the publication of *On the Origin of Species*.[163] In more recent times, the evangelical Anglican church leader John Stott provides a good example of this intermediate position. In particular, he insisted that it was possible to believe in a historical Adam and Eve and a historical fall, while at the same time considering that their creation in God's image may have come about through an evolutionary process:

---

[161]Wright, "A British Reflection on the Evolution Controversy in America," 137.

[162]J. I. Packer, "Foreword," in *Darwinism Defeated?*, 7.

[163]Early responders to Darwin such as the Catholic biologist George Mivart, the Methodist geologist Alexander Winchell, and the Presbyterian biologist George Macloskie sought to accommodate his theory within a relatively conservative framework, maintaining a historical Adam and Eve and ultimately a monogenetic account of human origins. See the overview in David Livingstone, *Adam's Ancestors: Race, Religion, and the Politics of Human Origins* (Baltimore: Johns Hopkins University Press, 2008), 139-59. Note, however, that Mivart's views changed over his career.

> It seems perfectly possible to reconcile the historicity of Adam with at least some (theistic) evolutionary theory. Many biblical Christians in fact do so, believing them to be not entirely incompatible. To assert the historicity of an original pair who sinned through disobedience is one thing; it is quite another to deny all evolution and assert the separate and special creation of everything, including both subhuman creatures and Adam's body. The suggestion (for it is no more than this) does not seem to me to be against Scripture and therefore impossible that when God made man in His own image, what He did was to stamp His own likeness on one of the many "hominids" which appear to have been living at the time.[164]

In his *Understanding the Bible*, Stott further defended his view that some forms of evolutionary theory can be reconciled to the Christian faith:

> I cannot see that at least some forms of the theory of evolution contradict or are contradicted by the Genesis account of creation. It is most unfortunate that some who debate this issue begin by assuming that the words "creation" and "evolution" are mutually exclusive. If everything has come into existence through evolution, they say, then biblical creation has been disproved, whereas if God has created all things, then evolution must be false. It is, rather, this naïve alternative which is false. It presupposes a very narrow definition of the two terms.[165]

In continuing his reflections on Adam and Eve, Stott coined the now-famous term *homo divinus* as a reference to a human being who is made in the image of God:

> My acceptance of Adam and Eve as historical is not incompatible with my belief that several forms of pre-Adamic "hominid" seem to have existed for thousands of years previously. These hominids began to advance culturally. They made their cave drawings and buried their dead. It is conceivable that God created Adam out of one of them. You

---

[164]John Stott, *The Church of England Newspaper*, June 17, 1968, as quoted by N. M. de S. Cameron, *Evolution and the Authority of the Bible* (Exeter: Paternoster, 1983), 63.

[165]John Stott, *Understanding the Bible* (1972; rev. ed., Grand Rapids: Zondervan, 1999), 54-55.

> may call them *homo erectus*. I think you may even call some of them *homo sapiens*, for these are arbitrary scientific names. But Adam was the first *homo divinus*, if I may coin a phrase, the first man to whom may be given the biblical designation "made in the image of God."[166]

Stott was influenced by his contemporary and fellow Anglican Derek Kidner, who famously advanced the speculation that perhaps Adam was essentially a refurbished hominid while Eve was a de novo creation—a suggestion that Tim Keller has drawn attention to as a possibility.[167] The motivation for this hybrid view is not simply to accommodate evolutionary science but, as Keller observes, to make sense of some of the biblical data:

> This approach would explain perennially difficult Biblical questions such as—who were the people that Cain feared would slay him in revenge for the murder of Abel (Gen 4:14)? Who was Cain's wife, and how could Cain have built a city filled with inhabitants (Gen 4:17)? We might even ask why Genesis 2:20 hints that Adam went on a search to "find" a spouse if there were only animals around? In Kidner's approach, Adam and Eve were not alone in the world, and that answers all these questions.[168]

The notion that the *Adam and evolution* hypotheses might actually help explain some of the peculiarities of Genesis 1–4 will likely provoke those who regard such readings as constituting the abandonment of a historical reading of Genesis. But we must be wary of too simple a choice between historicity and symbol. Many conservative exegetes affirm the historicity of the text while recognizing a symbolical, pictorial thrust to much of its language. Henri Blocher, for instance, describes Genesis 1–4, in distinction from

---

[166]Stott, *Understanding the Bible*, 55-56.

[167]Derek Kidner, *Genesis: An Introduction and Commentary* (Downers Grove, IL: InterVarsity Press, 1967), 28-31.

[168]Tim Keller, "Creation, Evolution, and Christian Laypeople," *BioLogos*, February 23, 2012, https://biologos.org/articles/creation-evolution-and-christian-laypeople.

"straightforward, ordinary history," as "another historical genre: that of a well-crafted, child-like drawing of the far-distant past, with illustrative and typological interests uppermost—something like the images carved on the tympans of Romanesque cathedrals and the stories told by their stained-glass windows."[169] At the same time, Blocher's resistance to ahistorical, revisionist treatments of original sin is vigorous.[170] Similarly, J. I. Packer distinguishes the "poetic-prose mode of narration in Genesis 1–11, with its pictorial, imaginative, quasi-liturgical phraseology, its paucity of mere information, and its drumbeat formulae" from the "ordinary narrative prose mode" of Genesis 12–50. At the same time, Packer rejects the labels "legend, saga, epic, myth, or tale" to describe Genesis 1–11, affirming them as "archetypal history" and as "space-time history, although told in [Moses's] chosen incantatory-poetic way."[171] Elsewhere he speaks of the events of the early chapters of Genesis as historical events, but "shrouded in the mists of antiquity," and communicated in a subtle and secret manner.[172] Even those who are less eager to correlate Genesis 2–3 with questions of historical origins, such as the German Lutheran theologian Dietrich Bonhoeffer, are often misunderstood if our only interest is simply ascertaining whether they affirm or deny the text's historicity. Bonhoeffer spoke of Adam's placement in the garden, the two trees, and the other details of Genesis 2 as "ancient, magical pictures," and cautioned against the assumption that God cannot teach us through such mediums: "Who can speak of these things except in pictures? Pictures are not lies: they denote things, they let the things that are meant shine through."[173] For Bonhoeffer,

[169]Blocher, *Original Sin*, 41.

[170]E.g., Blocher, *Original Sin*, 56-62; cf. also Blocher, "The Theology of the Fall and the Origins of Evil," in *Darwin, Creation, and the Fall*, 159.

[171]J. I. Packer, "Hermeneutics and Genesis 1-11," *Southwestern Journal of Theology* 44.1 (2001): 13. I am indebted to Jack Collins for directing me to this work.

[172]J. I. Packer, "Foreword," in Melvin Tinker, *Reclaiming Genesis* (Oxford: Monarch, 2010), 13-14.

[173]Dietrich Bonhoeffer, *Creation and Fall; Temptation: Two Biblical Studies*, trans. John C. Fletcher (New York: Macmillan, 1959), 49-50.

Genesis 2–3 is telling the true story of who we are in the most appropriate way that this story can be told. Even those of us who might differ with Bonhoeffer may perhaps appreciate how his emphasis on the pictorial referentiality of Genesis 2–3 can introduce new complexities to the range of questions we ask of this text.[174]

These challenges are not exclusively post-Darwinian. Even prior to Darwin, interpreters wrestled with how to understand human origins and development in the early chapters of Genesis. The challenge of Cain and incest has especially loomed large. In 1856, three years before Darwin's *On the Origin of Species* appeared, the prominent British Orientalist and lexicographer Edward William Lane generated an enormous controversy over his supposal that "pre-Adamites" continued in existence after Adam and intermarried with the Adamic line. A significant feature of Lane's argument concerned Cain in Genesis 4. Quoting Genesis 4:14-17, he claimed:

> Cain may be supposed to have expected the great increase of Adam's posterity which happened during his life-time, and thence to have feared the vengeance of a kinsman: but this is certainly not the obvious meaning of his words: and moreover, he was, on the day of his saying thus, "driven out from the face of the earth" [or "land"], evidently meaning the land of his parents, and became "a fugitive and a vagabond." How then, should he fear the vengeance of his own kindred? His wife is commonly supposed to have been his sister: and at least one of Adam's sons must have married his sister if no other human race but that of Adam existed: but this is contrary to an express law of God.[175]

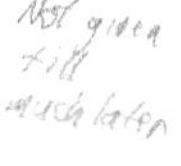

The hypothesis of "pre-Adamites" goes all the way back to the seventeenth century, when Isaac La Peyrère published his infamous

[174]In his exposition of Adam's formation in Genesis 2:7, Bonhoeffer stipulated that this imagery conveys God's nearness to, and authority over, human beings, and is not a literal account of how human beings were created: "This can surely not produce any knowledge about the origin of man!" (*Creation and Fall*, 46).

[175]Edward William Lane, *The Genesis of Earth and of Man*, ed. Reginald Stuart Poole, 2nd ed. (London: Williams and Norgate, 1860), 65-66. The first edition was published anonymously but is generally attributed to Lane. The editor, Reginald Poole, was Lane's cousin.

Prae-Adamitae (*Men Before Adam*) in 1655. A lawyer by profession, Peyrère was also an authoritative cartographer of northern lands such as Iceland and Greenland. His work in this area impressed upon him the challenge of establishing a biological relationship between various native populations and a recent, Middle Eastern Adam, and he delivered a withering critique of Hugo Grotius's suggestion that the Inuit in Greenland were of Norwegian descent.[176] Nonetheless, La Peyrère primarily made his case on exegetical grounds, in particular arguing that the "law" in view in Romans 5:13 is that given to Adam, not Moses, thus requiring sin before Adam.[177]

Although La Peyrère's views were widely and vigorously denounced in the church, he had admirers as well.[178] For those sympathetic to his thesis, however, Genesis 4 was probably a bigger factor than La Peyrère's somewhat eccentric argumentation from Romans 5:13. For instance, it was the difficulty with the notion of Cain as "the third man in the world" that led the English Protestant Gerard Winstanley (1609–1676) to affirm that "there were men in the world before that time" (i.e., the time of Adam).[179] Another common emphasis was the association of Adam with Jewish ancestry specifically, rather than human ancestry generally. This is the point urged, for instance, in Charles Blount's 1693 *Oracles of Reason*: "Moses made [Adam] only to be the Father of the Iews, whilst others Hyperbolically make him to be the first Father of all Men." On the contrary, this text affirmed that "there were two creations of Man and Woman, and

---

[176]Livingstone, *Adam's Ancestors*, 30. I draw much from Livingstone for my overview of La Peyrère's views and influence.

[177]Livingstone, *Adam's Ancestors*, 33.

[178]La Peyrère continued to be a controversial figure in succeeding generations. Livingstone, *Adam's Ancestors*, 219-20, notes: "The spiritual fate of pre-adamism's most conspicuous early champion, Isaac La Peyrère, is encapsulated in the following piece of doggerel, composed as an epitaph and loosely translated from the French:

Here lies La Peyrère, first a good Israelite,
Then Hugenot [ ], Catholic, Pre-adamite,
Four religions he tried, till, perplexed by so many,
At eighty he died, and went off without any."

[179]Quoted in Livingstone, *Adam's Ancestors*, 42.

that *Adam* was not the first Man, nor *Eve* the first Woman, only the first of the Holy race."[180] In succeeding generations, a wide array of pre-Adamite and co-Adamite[181] hypotheses emerged, some of them involving strange notions of distinct lineages involving demonic or angelic influence,[182] and occasionally accompanied by disturbing racial implications.[183]

It is somewhat ironic that while many of those accommodating evolution have maintained some species of monogenesis, many anti-Darwinists have been open to the existence of non-Adamic people and civilizations.[184] Thus, as Livingstone summarizes, although evolutionary theory is often seen as eclipsing belief in Adam, "for a significant body of opinion the coming of evolution meant the birth, not the death, of Adam."[185]

---

[180]Quoted in Livingstone, *Adam's Ancestors*, 42.

[181]Nathaniel Lardner's (1684–1768) suggestion of co-Adamites was seized upon by many others as a safer alternative to the pre-Adamite hypothesis, since it allowed Adam to still be a representative head of all humanity at the time of the fall, if not the genealogical head (Livingstone, *Adam's Ancestors*, 44-46). Lardner himself ultimately declined this view.

[182]Francis Dobbs (1750–1811) affirmed two distinct races of humanity, one from Adam and the other (among whom Cain settled) from Satan (Livingstone, *Adam's Ancestors*, 43); in the other direction, Isabelle Duncan (1812–1878) posited a pre-Adamite race, completely discontinuous with the Adamic line, which eventually became angels (Livingstone, *Adam's Ancestors*, 87-91). Duncan's work was published anonymously as *Pre-Adamite Man* at the same time as Darwin's *On the Origin of Species* made its appearance in the winter of 1859–1860. It was quite popular and quickly went through several editions and has received attention from contemporary scientists like Stephen Jay Gould as an early and creative attempt to integrate pre-Adamism within a relatively conservative theological framework.

[183]Livingstone, *Adam's Ancestors*, 109-200.

[184]For instance, the fundamentalist leader R. A. Torrey argued that everything after Genesis 1:1 in the creation account refers not to the creation but rather the reconstruction of a world decimated by "the sin of some pre-Adamic race." Torrey speculated further, "It may be that these ancient civilizations which are being discovered in the vicinity of Nineveh and elsewhere may be the remains of the pre-Adamic race already mentioned. . . . No one need have the least fear of any discoveries that the archeologists may make; for if it should be found that there were early civilizations thousands of years before Christ, it would not come into conflict whatever with what the Bible really teaches about the antiquity of man, the Adamic race" (quoted in Livingstone, *Adam's Ancestors*, 202). Similarly, Ambrose Fleming, another critic of evolutionary thought, suggested that the Neanderthals were a pre-Adamic species in order to resist their discovery as an evidence for human evolution. He also identified the Cro-Magnons as the antediluvians in the Bible (Livingstone, *Adam's Ancestors*, 205).

[185]Livingstone, *Adam's Ancestors*, 137. Cf. 168: "Evolution, far from heralding the death of Adam, in fact confirmed the very opposite: the literal birth of Adam."

At the same time, we must recognize that harmonizing a historical Adam with evolutionary science is no straightforward task. It generates a host of thorny questions. One of the greatest challenges to the current proposals concerns *when* exactly Adam and Eve are situated within the evolutionary schema. A broad fault line lies between those who affirm a more recent (e.g., Neolithic) Adam and Eve, who function as representative or archetypal human beings, and those who affirm an older Adam and Eve, who functioned as, among other things, the biological progenitors of the entire human race.[186] We might call this latter option the *ancient Adam* view. Proposals differ, though, on what is meant by "ancient." Some want to date Adam around the origins of anatomically modern *Homo sapiens* (sometime between 100,000–200,000 years ago),[187] while others want to push him all the way back to somewhere around the emergence of the genus *Homo* (roughly 2 million years ago), pointing to morphological discontinuities with the australopithecines.[188] These views are often motivated by the laudable concern to protect the unity of humankind—by placing Adam further back, within a relatively smaller population within a relatively smaller geographical region, it is easier to envision how the spread of sin and death to all humanity might have occurred.[189]

---

[186]Some of the material that follows was originally developed in response to Doug Moo's paper on the historical Adam in Paul at the Dabar Conference at Trinity Evangelical Divinity School, June 2018.

[187]This view is represented, for instance, by the old-earth creationist ministry Reasons to Believe, a revision of their earlier hypothesis of roughly 50,000–60,000 years ago. See Fuz Rana with Hugh Ross, *Who Was Adam? A Creation Model Approach to the Origin of Man* (Covina, CA: RTB Press, 2015), 267.

[188]As advocated by the anonymous scientist under the pseudonym William Stone, "Adam and Historical Science," in *Adam, the Fall, and Original Sin,* 80. The australopithecines are a genus of extinct early hominins that lived 2–4 million years ago, just prior to the emergence of genus *Homo*.

[189]This is somewhat dependent, of course, on what the precise role of Adam and Eve is construed to be—for instance, some proposals in this camp conceive of Adam and Eve as our common progenitors, but not our sole progenitors. For a recent exposition of an ancient Adam view that addresses this distinction, see Michael Murray and Jeffrey Schloss, "Evolution," in *The Routledge Companion to Theism*, ed. Charles Taliaferro, Victoria Harrison, and Stewart Goetz (London: Routledge, 2012), 227-28.

However, there are challenges for the ancient Adam proposal. First, this view typically requires an African location for Adam and Eve, sometime before the emigration of *Homo sapiens* to other continents. Some have argued that features of Genesis 2–4 suggest, by contrast, a Near Eastern setting (for instance, the reference to Eden as "in the east" in Gen 2:8, and the identification of the Tigris and Euphrates Rivers) as well as a Neolithic timeframe (for instance, references to farming in Gen 4:12, precious metals such as gold and onyx in Gen 2:11-12 and bronze and iron in Gen 4:22, and musical instruments such as the lyre and pipe in Gen 4:21).[190] Second, while it is widely acknowledged that biblical genealogies may contain gaps and cannot be used for exact dating, it is questionable whether they can be stretched out as far as these proposals require.[191] Finally, positing an older Adam must account for those interbreeding events between *Homo sapiens* and both Neanderthals and Denisovans that allegedly occurred up to about 30,000 years ago, and the resultant theological difficulties. For instance, suppose a sinful, image-bearing descendent of Adam interbreeds with a Neanderthal; do their offspring bear the image of God?[192] This dilemma holds unless we posit that other species within the genus *Homo* are image bearers and possess moral accountability. As the anonymous scientist under the pseudonym William Stone (who himself holds this view) acknowledges, this entails the uncomfortable prospect that "Adam's progeny split into different species."[193]

[190]Denis Alexander, *Creation or Evolution: Do We Have to Choose?* 2nd ed. (Grand Rapids: Monarch, 2014), 301; R. J. Berry, "Did Darwin Dethrone Humankind?" in *Darwin, Creation, and the Fall*, 63n81.

[191]As observed by Deborah B. Haarsma and Loren D. Haarsma, *Origins: Christian Perspectives on Creation, Evolution, and Intelligent Design* (Grand Rapids: Faith Alive, 2011), 260: "The genealogies in Genesis may have skipped some generations, but it seems unlikely that those skips could cover a gap of 142,000 years." Throughout this chapter the Haarsmas provide a helpful taxonomy of five possible ways to harmonize Adam and Eve with evolution (*Origins*, 251-73).

[192]This challenge is anticipated by Stone, "Adam and Historical Science," in *Adam, the Fall, and Original Sin*, 62.

[193]Stone, "Adam and Historical Science," in *Adam, the Fall, and Original Sin*, 78.

In the other direction, we might consider a more recent, Neolithic Adam, not as the biological progenitor of all modern humans, but in federalist terms as a kind of representative head of humanity.[194] But this *recent Adam view* faces challenges as well. For instance, it must find some way to account for the transmission of the *imago Dei* and sinfulness from Adam and Eve to those who had already emigrated to Australia (approximately 50,000 years ago) and the American continents (approximately 10,000–20,000 years ago; sometimes estimated earlier). Some, perhaps, have no difficulty with this. R. J. Berry, for instance, supposes, "If [the *imago Dei*] was initially conferred on an individual, there is no reason why it should not spread by divine *fiat* to all other members of *Homo sapiens* alive at the time."[195] Similarly, Derek Kidner is content to hypothesize that "Adam's 'federal' headship of humanity extended . . . outwards to his contemporaries as well as onwards to his offspring, and his disobedience disinherited both alike."[196]

But when we conceptualize a "lateral" transmission of original sin, it is one thing to envision this occurring (for instance) from a chieftain to his tribe, and another to envision it over thousands of miles, multiple oceans, and significant cultural and linguistic barriers. Unless we embrace a model in which the full effects of the fall took time to spread outward to all humanity, this view runs the risk of requiring us to view Genesis 3 as some kind of *representation* of what humanity is and became, rather than an account of that change actually happening—akin to Denis Alexander's metaphorical comparison that the UN's Universal Declaration of Human Rights in 1948 did not cause human rights, but simply affirmed a preexisting reality.[197] While the story of Genesis 3 is pictorial and symbol laden, trading causation for description is a high price to pay, and a departure from traditional readings.

[194]Cf. Walton, "A Historical Adam," in *Four Views on The Historical Adam*, 89-118; and Walton, *The Lost World of Adam and Eve*.

[195]R. J. Berry, "Did Darwin Dethrone Humankind?" in *Darwin, Creation, and the Fall*, 66-67.

[196]Kidner, *Genesis*, 29.

[197]Alexander, *Creation or Evolution*, 296.

Perhaps the wisest approach, if we end up becoming convinced that we must accept both evolution and Adam, is simply to refrain from assigning a date to Adam at this point in time. It is possible, after all, to affirm a broadly traditional account of the theological role of Adam and Eve, recognize the plausibility of the data suggesting an evolutionary account of human origins, and simply be uncertain as to how these two bodies of knowledge relate to one another. Such an approach allows us to avoid a premature judgment about either the theology or the science, but instead to keep patiently working at both. It is not easy to live in this tension, but it is possible. There is still a considerable body of core doctrines within the doctrine of creation that can be delineated—for instance, one pastor has identified ten theses that comprise an account of "mere creation" that most evangelicals can agree on and can function in a church context as a set of boundary markers.[198] Or consider the carefulness reflected in the approach of C. John Collins, who offers a fairly traditional account of Adam and Eve—maintaining that the origin of humanity goes beyond a purely natural process, that Adam and Eve are the headwaters of the entire human race, and that the fall was a historical and moral event. At the same time, Collins allows that "even if someone is persuaded that humans had 'ancestors,' and that the human population has always been more than two, he does not *necessarily* have to ditch all traditional views of Adam and Eve."[199] Collins further envisions the possibility of a group of humans living at the same time as Adam and Eve as a "single tribe" of whom Adam is the "chieftain."[200] Collins surveys various other attempts to maintain a belief in a historical Adam and Eve alongside

[198]See Todd Wilson, "Mere Creation: Ten Theses (Most) Evangelicals Can (Mostly) Agree On," in *Creation and Doxology: The Beginning and End of God's Good World*, ed. Gerald Hiestand and Todd Wilson (Downers Grove, IL: IVP Academic, 2018), 45-58.

[199]Collins, *Did Adam and Eve Really Exist?*, 120, italics his. See also Collins, *Reading Genesis Well: Navigating History, Poetry, Science, and Truth in Genesis 1-11* (Grand Rapids: Zondervan, 2018).

[200]Collins, *Did Adam and Eve Really Exist?*, 121.

some kind of polygenesis,[201] though he specifies that his preferred view avoids this.[202]

In his endorsement of Collins's work, Henri Blocher describes Collins as "standing firm on vital issues, accepting diversity on others."[203] This balanced posture seems resonant with Augustine's penchant toward openhandedness on peripheral and detailed matters, on the one hand, and groundedness on central and weighty matters, on the other. Such an approach would be welcome in the current moment, as we continue to pursue the truth on this topic.

CONCLUSION

The conclusions of this chapter may be summarized in several theses. First, Augustine's conception of *rationes seminales* should check the assumption that gradual evolutionary processes are an inherently inferior means of creation. At the same time, it is a stretch to hitch this conception to biological evolution per se, and the way Augustine wields it undermines resistance to supernatural intrusion into the creative process. Augustine's vision of divine agency in creation may incline us to consider evolutionary mechanisms with more openness, but it by no means settles the issue in advance.

Second, Augustine sees a high level of symbolic and pictorial language in Genesis 2–3, and at times engages the story allegorically. Nonetheless, he insists on the essential historicity of the events recorded, and resists a dichotomy between the story's historical reference and its symbolic or spiritual implications. He strongly rebukes the error of too lightly leaving off a historical reading of Genesis 2–3. The indignation and vigor of his rebuke should not be lightly forgotten today.

---

[201] Collins, *Did Adam and Eve Really Exist?*, 105-31.

[202] See the interchange between Collins and Kevin DeYoung, "Adam and Eve Follow-Up: A Dialogue with Jack Collins," July 28, 2011, www.thegospelcoalition.org/blogs/kevin-deyoung/adam-and-eve-follow-up-a-dialogue-with-jack-collins.

[203] Collins, *Did Adam and Eve Really Exist?*, inside cover.

Third, Augustine retains a surprising degree of flexibility with respect to interpreting particular details in Genesis 2–3. Although he rejects a symbolical view of Adam and Eve and Eden, he does not regard this as an error that threatens orthodoxy; he even hedges his position to allow for its possibility. Moreover, he is open on details such as whether Adam was made as a man or child. He is more interested in the spiritual meaning of Eve's creation, and he approaches figuratively many other particular matters of interpretation, such as the nature of the curses in Genesis 3.

All this suggests that Augustine would encourage caution in assessing the relation of Adam and Eve to evolutionary claims. While Augustine can resource those who want to maintain a historical fall, he should not be used to buttress an entrenched position that simply skates past the scientific challenges without critical engagement, closed off to any possible revision or further nuance. Much in Augustine's theology is favorable to harmonization efforts in the realm of instinct three (*Adam and evolution*).

CONCLUSION

RECAPPING AUGUSTINE'S INFLUENCE ON THE CREATION DEBATE

WE BEGAN THIS BOOK BY PICTURING AUGUSTINE joining our table conversation. Since conversations are often edifying in ways that are difficult to quantify, it may not be possible to articulate everything we have learned from Augustine. Nonetheless, it may be useful to try to sum up some of the important points of his treatment of creation, particularly as they relate to our engagement with this doctrine today. If Augustine were at the table, how might the conversation about creation move differently? Here I simply reiterate one representative lesson from each chapter.

First, Augustine helps us wonder at sheer createdness. Creation is not a necessity. It reflects the generosity of God. As Augustine prays, "You created, not because you had need, but out of the abundance of your own goodness."[1] Related to this, creation is, for Augustine, an emotional doctrine. He engages it at a deeply existential level. Specifically, he holds that the human soul was made for God, and thus every facet of human existence is dynamically oriented toward God. At every moment and in all that we do, we are constantly upheld by God, relating to God, and in need of God. He is the constant fact with which all existence has to do. Life and happiness are fully and only

[1] *Confessiones* 13.4 (CSEL 33, 348).

from him. As Augustine writes, "Even when all is well with me, what am I but a creature suckled on your milk and feeding on yourself, the food that never perishes?"[2]

Augustine's vision of creaturely dependence on God extends not only to the human soul but to the entire created universe. The whole world is reverberating with imperfection, longing to share with the angels in divine immutability—like a piece of pottery that has been constructed but has not yet gone through the firing and glazing stages. It awaits its final confirmation in God. Augustine may as well have prayed, "You have made all creation for yourself, and it will find no rest until it rests in you." This is why the *Confessions* ends with Genesis 1, and why Genesis 1:1-2 sets Augustine's heart throbbing.[3]

Augustine's reminder of the miracle of creatureliness makes it more difficult to take this doctrine for granted, or to put all our focus on simply *how* it happened. Creation is not a speculative topic but vitally concerns human happiness. Even secular people involved in the conversation may be intrigued by Augustine's insights into the craving of the human heart.

Second, Augustine's humility concerning the doctrine of creation encourages irenicism, particularly in the relation of theology and (what we would call) science. Now, this is not to say that Augustine is unwilling to debate about creation. He is deeply concerned to affirm the goodness of creation, for example, in response to Manichaean errors. We feel from Augustine the importance of creation and its foundational significance for the doctrines of sin and redemption. He is always willing to reject overreaching claims from philosophers, particularly when they threaten orthodoxy.

At the same time, within the rule of faith, Augustine is remarkably circumspect. His great concern is to avoid rashness (*temeritas*). He has enormous respect for the work of *philosophi* and *medici*, and is

---

[2]*Confessiones* 4.1 (CSEL 33, 64).

[3]*Confessiones* 12.1 (CSEL 33, 310).

horrified at anti-intellectual dismissals of genuine discovery. He insists on the complete trustworthiness of Scripture, but remains keenly alert to his fallibility as an interpreter. He works hard to harmonize biblical texts with each other and with other fields of knowledge. He often functions with approximate or provisional views. He is willing to reconsider his claims.

Augustine's presence in the current creation debate would encourage a more complicated view of the relation of Scripture and science, and more care in relating them. I can well imagine Augustine at the table, holding up his hands in protest, urging caution, listening, and patience—or, to use his terms, calling for less "obstinate wrangling" and more "diligent seeking, humble asking, persistent knocking."[4]

In both of these first two points—Augustine's expansive vision of creation and his humble method of engaging it—he may remind us that the most important aspects of the doctrine of creation are not those typically disputed among Christians, but those held in common (such as creation *ex nihilo*). To put it colloquially, he might help us major on the majors, and minor on the minors.

Third, Augustine helps us to appreciate the complexity of interpreting Genesis 1. Having felt Augustine's anxiety over this passage, and having traced the development of his views throughout his five commentaries, it will be more difficult to rebuke all those who can't see its obvious "plain meaning." Augustine may prompt us to deeper hermeneutical considerations when he suggests that the days function as an act of divine accommodation, "as a help to human frailty . . . to suggest sublime things to lowly people in a lowly manner by following the basic rule of story-telling."[5] He will certainly complicate our terminology, since he regards a "literal" interpretation of Genesis 1 as concerning historical referentiality without excluding allegorical

[4]*De Genesi ad litteram* 10.23.39 (CSEL 28:1, 326).
[5]*De Genesi ad litteram liber unus imperfectus* 3.8 (CSEL 28:1, 463).

meaning or various kinds of figurative language. Some of his views on Genesis 1 may prompt quizzical looks, like his claim that the ordering of events is according to angelic knowledge. Yet the influence of Augustine's exegesis of Genesis 1, particularly through the medieval era, discourages us from simply writing him off as an eccentric. Recall that Andrew Brown calls Augustine's interpretation "the defining statement with which every medieval and Renaissance commentator on Gen. 1:1–2:3 would wrestle."[6]

In the current creation debate, the vigor of Augustine's rejection of twenty-four-hour days will certainly be felt. For Augustine, "it can scarcely be supposed" that light turned on and off on days one to three before the creation of the sun;[7] it is "beyond a shadow of doubt" and "limpidly clear" that Genesis 2:4-6 confirm non-ordinary days;[8] it is "the height of folly" to read day seven in a literalistic way.[9] If all this fails to convince, there remains the challenge of squeezing the events of day five into twenty-four hours: "Here, surely, anyone slow on the uptake should finally wake up to understanding what sort of days are being counted here."[10]

The force of these rebukes is partially explained by the fact that Augustine associates literalism with the Manichaeans. Yet his views still undermine the claim that all rejection of twenty-four-hour days in Genesis 1 is motivated by scientific discovery. If we appeal to twenty-four-hour days as the "plain reading," we must reckon with the towering fact that the greatest theologian of the early church found the opposing view equally "plain." This would seem, at the very least, to encourage more space for legitimate disagreement concerning the interpretation of Genesis 1.

---

[6]Andrew J. Brown, *The Days of Creation: A History of Christian Interpretation of Genesis 1:1–2:3*, History of Biblical Interpretation (Blandford Forum, UK: Deo, 2014), 53.

[7]*De Genesi ad litteram* 1.11.23 (CSEL 28:1, 17).

[8]*De Genesi ad litteram* 5.1.3 (CSEL 28:1, 138).

[9]*De Genesi ad litteram* 4.8.16 (CSEL 28:1, 104).

[10]*De Genesi ad litteram liber unus imperfectus* 15.51 (CSEL 28:1, 495).

Fourth, Augustine represents a different set of intuitions about animal death. His fascination with insects is particularly unforgettable. He marvels at the rhythmic movement of worms,[11] the agile flight of flies,[12] the tiny labors of ants.[13] God made all these creatures.[14] They reflect his wisdom.[15] We should praise God for them.[16] They teach us valuable spiritual lessons.[17]

Here again it is not merely the content of his views that might instruct us, but the vigor with which he holds them. He not only defends the predatory system of animal life as a kind of beauty; he thinks it is "ridiculous"[18] and "foolish"[19] and "the last word in absurdity"[20] to sit in judgment on it. To blame God for creating fearsome animals is like a layman judging the mechanic's instruments because he has used them improperly.[21] We should not, Augustine insists, make such self-referential judgments. Instead, we should remember that our perspective is limited. God's work of creation is like a massive poem, such that those who object to the passing away of some creatures and cannot appreciate the larger purposes of divine providence are "behaving as absurdly as if someone in the recitation of some well-known poem wanted to listen to just one single syllable all the time."[22] The fact that Augustine's rebukes must be understood in context as directed toward the Manichaeans does not drain them of contemporary relevance.

---

[11]*De vera religione* 41.77 (CSEL 77, 56).

[12]*De Genesi ad litteram* 3.14.22 (CSEL 28:1, 79-80).

[13]*De Genesi contra Manichaeos* 1.16.26 (CSEL 91, 93-94).

[14]*De Genesi ad litteram* 3.14.22 (CSEL 28:1, 79).

[15]Sermon 29.7, in *Sermons Discovered Since 1990*, ed. John E. Rotelle, trans. Edmund Hill, The Works of Saint Augustine: A Translation for the 21st Century (Hyde Park, NY: New City Press, 1997), 58

[16]*Confessiones* 10.35 (CSEL 33, 269-70).

[17]*De Genesi ad litteram* 3.16.25 (CSEL 28:1, 82).

[18]*De civitate Dei* 12.4 (CSEL 40.1, 571).

[19]*De libero arbitrio* 3.15 (CSEL 74, 125-26).

[20]Sermon 29.7, in *Sermons Discovered Since 1990*, 58.

[21]*De Genesi contra Manichaeos* 1.16.25 (CSEL 91, 91-92).

[22]*De vera religione* 22.43 (CSEL 77, 30).

Given the emotional and rhetorical volatility of this topic, Augustine's differing instincts may serve as a calming influence. And his particular insights—temporal beauty and perspectival prejudice—may be useful in pursuing an answer to the question of animal death.

Finally, Augustine may help us engage challenges posed by evolutionary science to the historicity of Adam and Eve as well as the fall. Augustine resists the tendency to choose between history and symbol in Genesis 2–3. He worries about literalism with respect to many textual details, such as: What is the nature of the "mud" in Genesis 2:7, and how did God "fashion" Adam from it?[23] How precisely did Satan "use" the serpent?[24] What are the prophetical aspects of the curses of Genesis 3?[25] He often appeals to angelic agency to protect against the dangers of literalism.[26] At the same time, he maintains that whatever literary complexities we may find in Genesis 2–3, these chapters are inextricably concerned with the origins and fall of humanity in real history.[27]

Augustine affirms a supernatural creation of Adam as the first human being, but leaves it open whether Adam was created as an infant or adult.[28] For Augustine, those who affirm a merely symbolical garden of Eden and Adam are wrong but still orthodox; unlike the Manichaeans, they are fellow Christians who revere Scripture.[29] Augustine is sincerely worried about the implications of denying a real Adam.[30] But, if it became impossible to understand Adam as a

---

[23]Augustine worries about the "excessively childish notion" that "God molded the man from mud with actual material hands," while also resisting the Manichaean objection to the use of mud in the process (*De Genesi ad litteram* 6.12.20 [CSEL 28:1, 185]).

[24]*De Genesi ad litteram* 11.27.34 (CSEL 28:1, 360).

[25]For instance, as we observed, Augustine takes the curse on the woman in a prophetical sense (*De Genesi ad litteram* 11.37.50 [CSEL 28:1, 372]), while the curse on the man is both historical and prophetical (*De Genesi ad litteram* 11.38.51 [CSEL 28:1, 373]).

[26]For example, he appeals to "the ministry of angels" in the bringing of the animals to be named by Adam in order "to save us from being crassly literal-minded" (*De Genesi ad litteram* 9.14.24 [CSEL 28:1, 285]).

[27]E.g., *De civitate Dei* 13.21 (CSEL 40.1, 645) or *De Genesi ad litteram* 8.1.2 (CSEL 28:1, 230).

[28]*De Genesi ad litteram* 6.13.23 (CSEL 28:1, 187-88).

[29]*De Genesi ad litteram* 8.1.4 (CSEL 28:1, 231).

[30]*De Genesi ad litteram* 8.1.4 (CSEL 28:1, 231).

historical person, it would be necessary to interpret Genesis 2–3 figuratively rather than find fault with Holy Scripture.[31]

Such concessions about Adam will be astonishing to many participants in the current creation debate, given Augustine's conservative credentials. At the same time, we must consider Augustine's worry about leaving off historicity too glibly. Ultimately, Augustine's treatment of this topic would likely bolster the effort to maintain a historical Adam and Eve while encouraging openness in how the details of such an effort relate to current scientific claims. The flexibility of his posture toward this challenge might ease some of the anxiety we feel in the process.

Throughout this book, we have tried to listen carefully to Augustine's views; hopefully we have been enriched by them. At a deeper level, we may find that something of the spirit of Augustine's theology, something of his pathos and wonder, has gotten in our blood. Theology is, after all, more than our mere positions; it is a *coram Deo* task that involves our whole being. Thus, we may not only be grateful to Augustine for whatever progress we have made; we may discover that his voice stays with us for the work that remains ahead.

[31]*De Genesi ad litteram* 8.1.4 (CSEL 28:1, 232); cf. his earlier concession to this effect in *De Genesi ad litteram* 6.21.32 (CSEL 28:1, 195) and his latter comments in a similar vein with respect to the tree of the knowledge of good and evil in *De Genesi ad litteram* 11.41.56 (CSEL 28:1, 376).

GENERAL INDEX

SCRIPTURE INDEX